T0251380

Integrated Inductors and Transformers

Characterization, Design
and Modeling for RF and
mm-Wave Applications

Integrated Inductors and Transformers

Characterization, Design and Modeling for RF and mm-Wave Applications

Angelo Scuderi • Egidio Ragonese
Tonio Biondi • Giuseppe Palmisano

CRC Press
Taylor & Francis Group
Boca Raton London New York

CRC Press is an imprint of the
Taylor & Francis Group, an **informa** business
AN AUERBACH BOOK

Auerbach Publications
Taylor & Francis Group
6000 Broken Sound Parkway NW, Suite 300
Boca Raton, FL 33487-2742

International Standard Book Number: 978-1-4200-8844-1 (Hardback)

Library of Congress Cataloging-in-Publication Data

Integrated inductors and transformers : characterization, design and modeling for RF
    and MM-wave applications / Egidio Ragonese ... [et al.].
        p. cm.
    Includes bibliographical references and index.
    ISBN 978-1-4200-8844-1 (hardback)
    1. Radio frequency integrated circuits. 2. Electric inductors. 3. Electric transformers.
I. Ragonese, Egidio. II. Title.

TK7874.78.I58 2010
621.3815--dc22   2010035472

Visit the Taylor & Francis Web site at
http://www.taylorandfrancis.com

and the Auerbach Web site at
http://www.auerbach-publications.com

Contents

List of Figures

List of Tables

1

INTRODUCTION

The history of electromagnetism saw the light with the early studies of Michael Faraday in 1821 and James Clerk Maxwell in 1865, who first theorized the existence of electric and magnetic waves. However, experimental evidence of Maxwell theories was found several years later thanks to the research carried out by David E. Hughes in 1879, although wrongly attributed to magnetic induction rather than electromagnetism; Thomas A. Edison, who obtained the very first wireless patent of the history in 1885; Heinrich Hertz, who first demonstrated the possibility to transmit and receive electromagnetic waves in 1887; and Guglielmo Marconi, who realized the first long-distance transmission and reception of electromagnetic waves covering more than 3500 km over the Atlantic Ocean in 1901. At that time, neither Marconi nor any of his predecessors could ever be aware of what such discoveries would have given beginning to. Wireless communications, as exchange of information over short or long distances in the form of acoustic waves, radio waves, or light (either visible or not), traveling in the ether, stimulated the scientific community to investigate this new frontier in all possible applications. But only in the 1960s, after prototyping the first silicon transistor in 1954 and when mass production of semiconductor transistors became practical, did all the exciting prospective become facts. Obviously first applications came from the military field, where wireless communications represented a big attractive, whereas for a long time civil applications of the ether transmission were limited to radio and television broadcasting. At that time, the bipolar junction transistor (BJT) was the most commonly used transistor. Even after the metal oxide semiconductor field-effect transistor (MOSFET) became available, the BJT remained the transistor of choice for many analog circuits because of its superior electrical properties and ease of manufacture.

Thirty years later, in the early 1990s, silicon transistors exceeded the 10-GHz unity-gain frequency frontier promoting silicon technology as a cost-effective platform for high-speed applications. The progress of fabrication technology and continuous shrinking of device size also allowed higher and higher level of integration to be achieved, exceeding the barrier of 1 million transistors in a chip. Pioneering examples of integrated inductors and *LC* filters were first proposed (Nguyen and Meyer 1990). This allowed the fabrication of the high-frequency wireless front-end together with the high-density low-power baseband and digital parts in a single chip at reasonable costs. More recently the prospective of a low-cost technology suitable for high-volume production, which had already fostered the conversion of consolidated wired communication systems, such as telephone or telegraph, to their wireless counterparts, also promoted the growth of new wireless standards featuring higher data rates, namely Bluetooth and ultra-wideband.

Today the scenario of wireless technology includes applications such as radio frequency identification (RFID), which is replacing the older barcode system as a security measure against shoplifting, wireless video surveillance systems for home or office buildings, television remote control, satellite digital video broadcasting, third-generation mobile cellular phones and modems, wireless local area networks (WLAN or WiFi), with hot spots located in most public buildings to enable internet connection of laptop PCs, personal digital assistants, and other devices, wireless energy transfer, and wireless computer interface devices. Thanks to the aggressive size scaling, nowadays silicon transistors have largely surpassed the 200-GHz unity-gain frequency frontier (Chevalier et al. 2005), opening the way for many other applications to be investigated in the automotive, industrial, medical, security, and domotics segments.

Figure 1.1 provides an overview of the operating bands allocated for the main applications in the radio frequency (RF) and millimeter wave (mm-wave) ranges.

The challenge of modern wireless communications is to implement competitive solutions in terms of integration, performance, and cost. A mass market viewpoint calls for the adoption of low-cost technologies capable of implementing complex functions. Because a transceiver is composed of the RF front-end, baseband, and digital parts, the choice

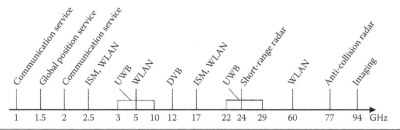

Figure 1.1 Frequency allocation of the main wireless applications in the RF and mm-wave ranges.

of an appropriate fabrication technology is one of the most critical issues. Depending on the application, the high-frequency part could require low noise, high linearity, and power capabilities that can be supplied by bipolar technologies. On the contrary, only deep submicron complementary metal oxide semiconductor (CMOS) processes can be used for the baseband and digital parts, which require high speed, high density, and low power consumption. Therefore, bipolar complementary metal oxide semiconductor (BiCMOS) technologies seem the obvious solution complying with all these matters (Long 2005). However, in the last few years high-frequency capabilities were also demonstrated by state-of-the-art CMOS technologies (Mehrotra et al. 2002), whose potential is already consolidated for application up to 5 GHz (Lee, Samavati, and Rategh 2002) but also investigated up to 60 GHz (Doan et al. 2005). State-of-the-art 60-nm SiGe BiCMOS and 45-nm CMOS technologies are capable of providing adequate high-frequency performance (Bennett et al. 2005; Joseph et al. 2005); however, they cannot be considered cost-effective solutions for mass market production due to their still high fabrication costs. On the other hand, alternative approaches to the monolithic integration of both the high- and low-frequency parts of a wireless transceiver exist. Besides the aggressive scaling of transistor dimensions, several efforts are now focused on assembly technologies using less expensive and more efficient packaging techniques. Advanced assembly techniques allow system-in-package (SiP) or system-on-chip (SoC) to be achieved. The multichip module (MCM) approach allows RF components and integrated circuits (ICs), each fabricated in a cost-competitive technology, to be assembled together, thus reducing size, cost, and complexity. Consequently, a low-cost silicon process could be used to fabricate the RF front-end and combined with baseband and digital blocks fabricated in a suitable CMOS technology. Moreover, duplex, filters, and even antenna can be fabricated exploiting MCM passive devices. This approach allows more flexibility and (probably) lower costs with respect to its monolithic counterpart, thus emerging as the more convenient solution for complex systems in the RF and mm-wave frequency ranges.

According to the above considerations, fundamental building blocks of the wireless transceiver, such as low-noise amplifiers, mixers, voltage-controlled oscillators, and even power amplifiers, are subjects of advanced studies in view of this radical technology change. However, the possibility to fully exploit low-cost technologies to meet the stringent requirements of the wireless market in the multigigahertz frequency range is subject to the availability of high-quality passive components. In fact, integrated inductors, transformers, and capacitors are widely used in RF and mm-wave ICs to implement impedance and noise-matching networks, interstage filters, differential-to-single-ended conversion, bandwidth enhancement, resonant loads, and many other functions. Despite the efforts propounded during the last decade, silicon-integrated passive devices (especially inductors and transformers) still represent a bottleneck that prevents further performance improvement from being achieved at reasonable costs but, at the same time, an attractive research topic with high potential that promises vast enhancements in the near future. This motivation constitutes the *fil rouge* that links up the arguments discussed in this book.

The discussion starts with a very basic and general description of the behavior of inductive components, making a quick recall of the underlying physics to demonstrate why energy loss takes place and how it is related to the shape, structure, and materials employed for its fabrication. On-wafer measurements and de-embedding are briefly discussed to focus the attention on the problem of test pattern parasitics and its impact on the accuracy of high-frequency measurements. An overview of fabrication technology and modeling approaches completes this preliminary and preparatory part of the book. The subsequent survey on integrated inductors and transformers in silicon technology provides a more in-depth description of the performance trade-offs and optimization strategies that come up when dealing with very large-scale integration (VLSI) production. A reference technology is described to provide practical examples of passive device fabrication. Substrate optimization is discussed and supported by experimental measurements of actual devices to highlight its importance in the overall device performance trade-off. Lumped element equivalent circuit modeling is treated extensively, giving detailed explanations for each model equation and relating it to device geometry. Experimental measurements of actual devices are employed to validate the soundness of such models over frequency and layout parameters and to critically compare with the state of the art. General optimization strategies and design guidelines for RF and mm-wave circuits using integrated passive devices are outlined to provide practical examples of the concepts discussed in this book. The design of the main building blocks of three current wireless applications at 5, 17, and 24 GHz will be described focusing the attention on passive devices. The performance of these designs is verified experimentally and critically compared to what is achieved in the state of the art. A quick glance to more sophisticated technologies is given at the end of the book, where fabrication of passive devices on glass and plastic substrates is discussed. The performance of inductors fabricated on glass, plastic, and silicon is compared to highlight the benefits of high-isolation substrates and to assess the cost–performance trade-offs. The design of a 5-GHz transceiver for WLAN using the MCM approach is also reported to demonstrate a high-performance yet cost-competitive solution that exploits highly selective filters for sliding intermediate frequency (IF) superheterodyne RF front-end fabricated on glass and a low-cost silicon bipolar technology.

1.1 Organization

Chapter 2 provides general information that is necessary for a thorough understanding of the rest of this book. The background theory of monolithic passive devices built on a lossy substrate is first described. The main loss phenomena that take place in monolithic inductors and transformers—for example, ohmic losses, skin and proximity effects in the metal trace, electric and magnetic coupling to the substrate— are described in detail, together with the rudiments of the underlying physics. The basic inductor and transformer layout geometries and related parameters are also described. On-wafer measurements of passive devices are then discussed to introduce the concepts of system calibration, test pattern parasitics, and de-embedding that are of primary importance when dealing with high-frequency measurements of small passive devices. A five-step de-embedding procedure is described and compared to more traditional techniques to highlight the pros and cons. Finally, an overview of fabrication technologies is given with a focus on how to improve the performance of monolithic inductors and transformers. Numerical and circuit modeling approaches are also discussed.

After the general description of device physics, commonly employed techniques for performance optimization of monolithic inductors are investigated in Chapter 3. Their feasibility and cost-effectiveness are also analyzed and related to state-of-the-art fabrication processes. Improvements at metal level, such as metal shunting/thickening, to increase the low-frequency quality factor of the coil are reported. Special attention is devoted to techniques for minimization of substrate losses because they represent the most limiting factor for the performance of silicon devices operating in the gigahertz frequency range. Commonly employed solutions such as oxide thickening, ground shielding, postprocessing, and three-dimensional structures are evaluated. The performance improvements achievable thanks to these techniques and feasibility of application to VLSI production are weighted, taking into account the required fabrication costs. Practical application examples and experimental results of actual devices are also reported to allow the reader what-if and parametric analyses of process and layout variations.

A comprehensive review of the modeling techniques for inductors and transformers developed during the last years is also presented in Chapter 3 and Chapter 4, respectively. Numerical, distributed, and lumped modeling approaches are described to highlight their pros and cons in terms of accuracy, complexity, and simulation time. Because lumped models are by far the most widely employed within the designer community, an in-depth assessment of formulas and closed-form expressions to evaluate the contribution of each element of a compact model is reported. Theories and procedures for coil inductance, resistance, and capacitance calculation are reviewed and their accuracy evaluated by comparison with on-wafer experimental measurements of actual devices. The equivalent circuit and employed expressions of lumped scalable models for inductors and transformers described in the literature are also reported. Trade-offs between accuracy and complexity are examined to identify the range of frequencies and layout parameters in which they can be safely employed. The accuracy of the inductor and transformer models are demonstrated by comparison with experimental data of a wide set of fabricated devices.

Chapter 5 reports a summary of rule-of-thumb design guidelines and optimization strategies for inductors and transformers to be used in RF and mm-wave circuit blocks. Criteria for the choice of coil shape, layout parameters, and substrate configuration are given on the base of the performance observed from experimental measurements of reference devices and data reported in the literature. This chapter also contains three circuit design examples and related optimization approaches where these guidelines are put into practice. Different applications, such as 5-GHz WLAN, 17-GHz industrial, scientific, and medical (ISM) communication, 24-GHz automotive radar sensor, and various circuit blocks—that is, low noise amplifier, mixer, filter and matching network, voltage-controlled oscillator, and power stage—are discussed in order to highlight the main optimization strategies for both inductors and transformers. Chapter 6 discusses the fabrication of inductor and transformers on dielectric substrates. The quality of inductive components fabricated on glass, plastic, and silicon substrates is compared to highlight cost–performance tradeoffs and fields of application. A reference technology based on a glass substrate is first described. Benefits achievable thanks to the high level of substrate isolation are discussed based experimental data of reference devices. The design of an MCM for 5-GHz WLAN with passive components fabricated on glass is also reported to demonstrate the improvements achievable at the system level and to compare with its silicon-based counterpart. The fabrication of passive components on plastic substrates is also discussed. Experimental data of geometrically scaled inductors are reported to highlight their RF performance. A simple scalable model is also presented and main differences with respect to classical lumped models for silicon-integrated inductors are highlighted.

References

Bennett, H. S., R. Brederlow, J. C. Costa, et al., 2005, "Device and Technology Evolution for Si-Based RF Integrated Circuits," *IEEE Transactions on Electron Devices*, vol. 52, pp. 1235–1258.

Chevalier, P., C. Fellous, L. Rubaldo, et al., 2005, "230-GHz Self-Aligned SiGeC HBT for Optical and Millimeter-Wave Applications," *IEEE Journal of Solid-State Circuits*, vol. 40, pp. 2025–2034.

Doan, C. H., S. Emami, A. M. Niknejad, and R. W. Brodersen, 2005, "Millimeter-Wave CMOS Design," *IEEE Journal of Solid-State Circuits*, vol. 40, pp. 144–155.

Joseph, A. J., D. L. Harame, B. Jagannathan, et al., 2005, "Status and Direction of Communication Technologies — SiGe BiCMOS and RFCMOS," *Proceedings of the IEEE*, vol. 93, pp. 1539–1558.

Lee, T. H., H. Samavati, and H. R. Rategh, 2002, "5-GHz CMOS Wireless LANs," *IEEE Trans. Microwave Theory and Techniques*, vol. 50, pp. 268–280.

Long, J. R., 2005, "SiGe Radio Frequency ICs for Low-Power Portable Communication," *Proceedings of the IEEE*, vol. 93, pp. 1598–1623.

Mehrotra, M., J. Wu, A. Jain, et al., 2002, "60 nm Gate Length Dual-Vt CMOS for High Performance Applications," *IEEE Symposium on VLSI Technology*, pp. 124–125.

Nguyen, N. M., and R. G. Meyer, 1990, "Si IC-Compatible Inductors and LC Passive Filters," *IEEE Journal of Solid-State Circuits*, vol. 25, pp. 1028–1031.

2

BASIC CONCEPTS

This chapter provides the reader with the general information necessary for a thorough understanding of the rest of this book. The background theory of monolithic passive devices built on a lossy substrate is first described in Section 2.1. Special emphasis is given to the description of the main loss phenomena that take place in monolithic inductors and transformers providing the rudiments of the underlying physics. Section 2.2 reports an overview of the basic inductor layout geometries and related design parameters. On-wafer measurements of passive devices are discussed in Section 2.3 to introduce the concepts of system calibration, test pattern parasitics, and de-embedding that are of primary importance when dealing with high-frequency measurements of small passive devices. Finally, Section 2.4 offers an overview of state-of-the-art fabrication technologies, with a focus on how to improve the performance of monolithic inductors and transformers. Numerical and circuit modeling approaches are also discussed in Section 2.4.

2.1 Basic Definitions and Loss Mechanisms

The biggest challenge in the design of integrated inductive devices on silicon is minimizing losses. This, in turn, requires maximizing the quality factor (Q) for a given inductance value (L) or, equivalently, maximizing the stored electromagnetic energy (E_S) while minimizing the dissipated energy (E_L) in a cycle, as summarized by equation (2.1).

$$Q = 2\pi \cdot \frac{E_S}{E_L} \qquad (2.1)$$

Applying the above equation to the ideal case of an isolated coil, where the only cause of loss is the series resistance (R_S) of the metal winding, leads to the more widely known equation (2.2)

$$Q = 2\pi \cdot \frac{P_S \cdot T}{P_L \cdot T} = \omega \cdot \frac{\frac{1}{2}L \cdot I^2}{\frac{1}{2}R_S \cdot I^2} = \frac{\omega \cdot L}{R_S} \qquad (2.2)$$

where P_S and P_L are the stored and dissipated power, respectively; I is the root-mean-square (rms) current flowing through the coil; and T is the cycle (inverse of the signal frequency).

In circuit design, the inductance value is normally imposed by the application; therefore, the primary goal is to lay the inductor with the highest quality factor. Low-noise amplifier design is a special case where circuit optimization can be achieved by maximizing the ωQL product at the operating frequency, as will be discussed in Chapter 5. Several equations are reported in the literature to predict the inductance of a coil with reasonable accuracy starting from the main geometrical parameters. Because the inductance of a coil only depends on such geometrical parameters, the main limitations of these equations due to first-order approximations are overcome by adding correction factors derived from electromagnetic simulations. On the other hand, an accurate prediction of the quality factor is much more complex because the involved loss phenomena that take place in the coil and substrate must be taken into proper account. The lack of closed-form expressions to predict the quality factor of integrated inductors on a lossy substrate poses a problematic question to the RF designer: How to select the geometrical parameters to maximize the quality factor at a given frequency for a given inductance value? The only way to design a monolithic inductor with optimum performance is to understand the root causes of losses and exploit the methodologies to minimize their detrimental effects.

The main energy dissipation phenomena that affect the performance of inductive devices fabricated on a lossy substrate are schematically depicted in Figure 2.1. At the frequencies of interest to RF designers, losses occur in the metal layers that form the coil as well as in the conductive layers below the coil.

Equations (2.3) and (2.4), which form the basis of modern electromagnetic theory, show the relationship between magnetic and electric fields as described by Maxwell (1865, 1873). A time-varying magnetic field B generates an electric field E that opposes to the magnetic field

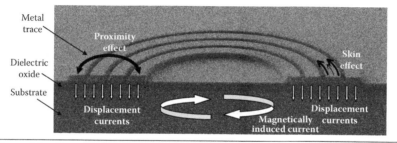

Figure 2.1   Energy dissipation phenomena of inductive devices on a lossy substrate.

itself (Faraday's law). Also, a time-varying electric field E or a current flow J, in turn, generates a magnetic field that supports the electric field itself (Ampere's law).

$$\nabla \times \vec{E} = -\frac{\partial \vec{B}}{\partial t} \tag{2.3}$$

$$\nabla \times \vec{B} = \mu \cdot \vec{J} + \mu \cdot \varepsilon \cdot \frac{\partial \vec{E}}{\partial t} \tag{2.4}$$

If applied in the context of this book, these equations can explain the root causes of power loss that take place in integrated inductors and transformers fabricated on lossy substrates. At this point, a distinction must be made between the so-called series losses, caused by the ohmic energy dissipation in the coil metal trace, and substrate losses, related to electromagnetic coupling between the coil and the semi-insulating material on which it was fabricated.

Series losses are caused by the finite conductivity of the metal layers used to fabricate the coil (copper or aluminum in VLSI technologies). At very low frequency, current flow is uniformly distributed inside the entire cross-sectional area of the conductor, which results in a constant-with-frequency coil series resistance. As the frequency increases (above few hundreds of kilohertz), current flow within the conductor is no more uniformly distributed but tends to crow toward its outer surface due to two distinct phenomena, referred to as *skin* and *proximity* effects. Both of them contribute to reduce the effective conductor cross-sectional area available for current conduction with respect to the DC case, which results in a coil-equivalent series resistance that increases with frequency. Skin effect takes place when an alternating current flows through an isolated conductor with finite conductivity. According to equation (2.4), current flow within the conductor generates a magnetic field lying in the plane orthogonal to that of the current flow. From equation (2.3), the generated time-varying magnetic field induces an electric field, lying in the same plane as the original current flow, that opposes to the magnetic field itself and, in turn, to the original current. Because the magnitude of the self-induced electric field is highest at the center of the conductor (decreasing along the radius), the current tends to crow toward the outer surface. This phenomenon is graphically explained in Figure 2.2(a) and Figure 2.2(b) for an isolated conductor with circular and rectangular cross section, respectively (the latter case is of concern for inductors and transformers fabricated in VLSI technology), where darker gray represents higher current density.

As the frequency increases—that is, as the rate of change of electric and magnetic fields increases—the magnitude of the self-induced electric field also increases, forcing the current to flow in a thin layer (skin) at the edge of the conductor cross-sectional area, which gives the name to this phenomenon. The resulting current density shows an exponential decay from the outer surface to the center of the conductor, as reported in equation (2.5)

$$J = J_0 \cdot e^{-\frac{x}{\delta}}, \ \delta = \sqrt{\frac{2 \cdot \rho}{\omega \cdot \mu}} \tag{2.5}$$

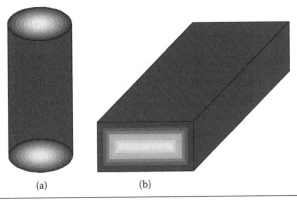

(a)   (b)

Figure 2.2   Skin effect in an isolated conductor with circular (a) and rectangular (b) cross section.

where x is the radial distance from the edge to the center of the conductor, δ is the penetration depth, ρ is the conductor resistivity, ω is the angular frequency, and μ is the magnetic permeability.

The other source of series losses is the proximity effect. It takes place when two or more conductors, where an alternating current flows, are placed close to each other (from whence the name). The causes of this phenomenon are very similar to those originating the skin effect and can thus be traced back using equations (2.3) and (2.4). However, the problem here is made more complex by the fact that the electric and magnetic fields generated by more conductors must be taken into account at the same time. Indeed, the time-varying current flow of each conductor generates a time-varying magnetic field and, in turn, an electric field in the conductor itself as well as in the neighboring ones. The current flow inside each conductor thus becomes nonuniform and tends to crow toward the outer surface. However, the current placement inside each conductor is more difficult to predict because it strongly depends on the spatial distribution of the neighboring conductors (how far are they and in which direction). If the above considerations are applied to a planar spiral, it comes out that current crowding is most pronounced in the innermost turns where the magnitude of the impressed magnetic field is highest. Figure 2.3 sketches the induced magnetic field inside a planar spiral due to proximity effect among the coils.

Besides series losses taking place within the coil metal trace, substrate losses contribute to worsen the overall quality factor of integrated inductors and transformers in silicon technology. The root cause of substrate losses is the electric and magnetic coupling between the metal layers of the spiral(s) and the underlying lossy substrate, which is again governed by equations (2.3) and (2.4). These losses come into play at higher frequencies than series losses; indeed, substrate conductivity is lower than metal trace conductivity, and therefore the generated time-varying electric and magnetic fields in the substrate have smaller magnitude at lower frequency. Due to the finite conductivity of the substrate, the generated electric and magnetic fields produce a flow of currents (so-called eddy currents) in the layers below the spiral(s). Two distinct phenomena play a role in the growth of these currents, as depicted in Figure 2.1. On one hand, the electrostatic coupling between the spiral and substrate causes vertical displacement

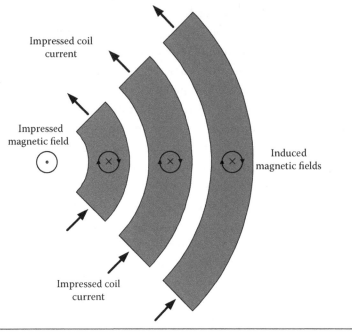

Figure 2.3 Proximity effect among the coils of a planar spiral.

currents through the dielectric layer (e.g., silicon oxide) that is supposed to isolate the spiral from the substrate. On the other hand, the
electromagnetic coupling due to the time-varying electric and magnetic fields causes ohmic currents in the substrate that flow antiparallel
to the impressed current as depicted in Figure 2.4. Besides increasing
the overall energy dissipation of the device, hence decreasing its quality factor, the flow of eddy currents in the substrate creates an induced
magnetic field that opposes the impressed one generated by the spiral
thus reducing the effective inductance of the coil by a small amount.

2.2   Layout Fundamentals

The basic structure for monolithic inductors and transformers is made
up of one or more metal windings placed on top of a bulk silicon
substrate with a dielectric insulator (silicon dioxide) in between. The
layout of the coil is defined by few geometrical parameters, namely,
the shape (square, hexagonal, octagonal, circular, etc.), the coil
trace metal width (w), the number of turns (n), the spacing between
adjacent turns (s), and the inner (d_{in}) and outer (d_{out}) diameters. Square

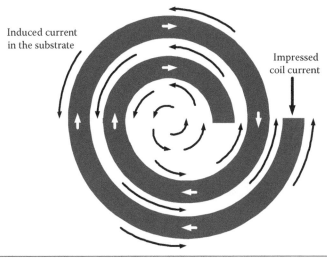

Figure 2.4 Eddy currents flow in the substrate below a planar spiral.

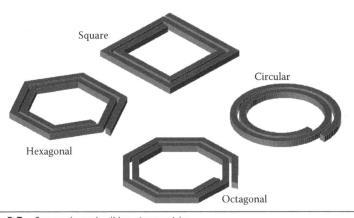

Figure 2.5 Commonly used coil layout geometries.

inductors were the most commonly used in the late 1990s, but they were soon after substituted by hexagonal, octagonal, and circular shapes that offer lower dc series resistance at a given inductance value. A three-dimensional view of the most commonly used layout geometries is reported in Figure 2.5.

These geometries are well suited for single-ended circuit topologies. However, many RF circuits are based on differential topologies to fully exploit virtual ground node benefits and common mode rejection. Although single-ended layout geometries might still be employed in differential topologies, they occupy a large amount of

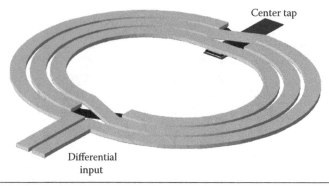

Figure 2.6 Symmetric coil geometry.

silicon area because two identical coils must be laid out to preserve symmetry. Dedicated geometries have thus been developed to achieve a better trade-off between symmetry and area occupancy in differential circuits, an example of which is given in Figure 2.6. A detailed description of the benefits of differentially driven passive devices will be given in Chapter 3.

2.3 Passive Device Measurements

Experimental measurements of on-wafer passive devices could be affected by high inaccuracy unless proper calibration and de-embedding techniques are employed (Biondi et al. 2007). This applies to inductor as well as resistor and capacitor measurements even if large devices are being measured. Indeed, the ground–signal–ground (GSG) structure commonly employed to perform on-wafer measurements at medium and high frequency offers parasitic paths to the test signal that might completely mask the measurement of the device under test (DUT) depending on both device size and measurement frequency. Indeed, as the device size shrinks down the impedances/admittances to be measured become comparable to the residual measurement parasitics that occur after system calibration. Residual errors deriving from the employed de-embedding technique also contribute to increase the overall relative error of measured data.

This section describes the techniques employed for achieving accurate high-frequency measurements of on-wafer passives devices. The precision of the measurement setup is validated taking into account both calibration and de-embedding issues. A five-step de-embedding technique is also reported and applied to the measurements of small inductors.

2.3.1 Calibration Accuracy and Residual Error

The measurement setup consists of a vector network analyzer (VNA) and a microprober for on-wafer measurements. Moving the measurement reference plane from the ports of the VNA to the probe tips can be accomplished using a set of on-wafer standard devices, referred to as *impedance standard substrate* (ISS), whose electrical behavior is accurately known over the frequency range of interest. Measured data of such standards are used to identify the unknowns of an error model that represents the measurement setup. This procedure, referred to as *calibration*, allows raw measurements collected by the VNA to be corrected from the nonideal behavior of the cables, adapters, and probes employed in the measurement system.

Among the calibration schemes employed in RF and microwave measurements, the line-reflect-reflect-match (LRRM) with automatic load inductance compensation was demonstrated to be the most accurate and repeatable up to very high frequency (Cascade Microtech 1994, 1998; Lord 2000). It fixes most of the disadvantages of previously employed calibration schemes, such as short-open-load-through (SOLT), short-open-load-reciprocal (SOLR), through-reflect-line (TRL), or line-reflect-match (LRM) (Davidson, Strid, and Jones 1989; Lautzenhiser, Davidson, and Jones 1990), without increasing the number of required standards. Moreover, it only involves the knowledge of the through-line delay and dc resistance of one load, thus avoiding the open and short to be accurately defined, as requested by the SOLT.

In order to quantify the amount of the residual error that occurs after calibration, the scattering parameters of the standard devices of the ISS can be measured. Measurements should be carried out and averaged on several devices to minimize the effect of the nonrepeatable contact impedance of the probes. The results of this are shown in Figures 2.7 through 2.9, where LRRM and SOLT (using the same number of calibration standards) are compared.

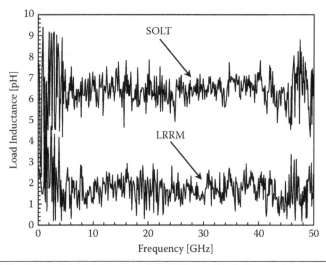

Figure 2.7   Measured load inductance after LRRM and SOLT. (From T. Biondi et al., 2007, "Characterization and Modeling of Silicon Integrated Spiral Inductors for High-Frequency Applications," *Analog Integrated Circuits and Signal Processing*, vol. 51, pp. 89–100. © 2007 Springer. With permission.)

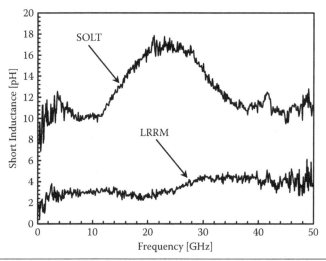

Figure 2.8   Measured short inductance after LRRM and SOLT. (From T. Biondi et al., 2007, "Characterization and Modeling of Silicon Integrated Spiral Inductors for High-Frequency Applications," *Analog Integrated Circuits and Signal Processing*, vol. 51, pp. 89–100. © 2007 Springer. With permission.)

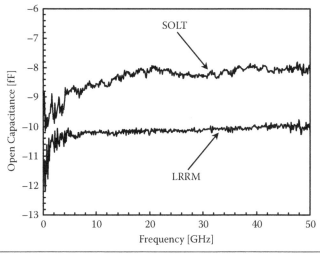

Figure 2.9   Measured open capacitance after LRRM and SOLT. (From T. Biondi et al., 2007, "Characterization and Modeling of Silicon Integrated Spiral Inductors for High-Frequency Applications," *Analog Integrated Circuits and Signal Processing*, vol. 51, pp. 89–100. © 2007 Springer. With permission.)

The load inductance extracted from the measured admittance (Y) parameters is shown in Figure 2.7. It can be observed that the residual inductance calculated with SOLT is much larger than that obtained with LRRM; indeed, averaging the measured data in the range 0.1–50 GHz provides a load inductance of 12.6 pH and 2.2 pH for SOLT and LRRM, respectively. Similar considerations can be drawn for the short inductance reported in Figure 2.8. In this case LRRM not only provides a lower average inductance compared to SOLT (3.5 pH instead of 12.9 pH) but shows much lower frequency dependence. Measurements of the open capacitance are reported in Figure 2.9. Because the probes lifted in air have less tip loading than when they are contacted to the ISS, the resulting capacitance has a negative value. Although the average capacitance obtained with LRRM (–10.2 fF) is higher (in absolute value) with respect to that of SOLT (–8.4 fF), the former value is closer to the value (–9.7 fF) reported for the 150-μm GSG ISS (Cascade Microtech 1994).

Besides demonstrating the soundness of the LRRM calibration technique experimentally, the above results estimate the resolution of the measurement system that can thus be used to obtain accurate measurements of small inductances. The maximum inductance deviation introduced by the (calibrated) system can be calculated as the maximum difference between the average load and short inductance obtained by measurements and the corresponding values reported in (Cascade Microtech 1994), which results in 3.9 pH. This translates into a maximum systematic error introduced by the measurement setup of 1% when measuring inductances as low as 0.39 nH.

2.3.2   De-Embedding Techniques

As the device size shrinks down, the parasitic effect of the pads and metal interconnects required to access the device terminals becomes increasingly dominant with respect to the electrical behavior of the device itself. Hence, careful de-embedding of measured data is of paramount importance especially at very high frequencies, where the impedance of even small parasitics might become comparable to that of the device under test. There are several ways to cancel out the effect of test pattern parasitics from raw measurements, whose degree of accuracy is strictly related to the device size, frequency range, and test pattern layout. In the past years, thanks to the relatively large size of the devices, removing the shunt parasitics of the pads using the open de-embedding method (Van Wijnen, Claessen, and Wolsheimer 1987; Frasen, Gleason, and Strid 1988) was adequate to achieve a reasonable accuracy up to medium frequencies. At higher frequencies, the effect of series parasitics due to the metal interconnects could not be neglected any more thus a three-step de-embedding method using on-wafer open, short, and through standards was proposed (Cho and Burk 1991). Improvements of the above technique were published more recently, extending its validity up to higher frequencies (Koolen, Geelen, and Versleijen 1991; Weng 1995; Kolding 2000; Vandamme, Schreurs, and van Dinther 2001).

A five-step de-embedding technique using five on-wafer test structures is now reported (Biondi et al. 2004). The impedance model of the de-embedding technique and related on-wafer test structures are sketched in Figures 2.10 and 2.11, respectively.

The correction procedure can be summarized as follows. The first step consists in taking into account the contact impedance of the probe tips by subtracting the impedance (Z) parameters of the pro-short from those of the DUT (and all remaining test structures) according to equation (2.6).

$$Z_{DUT}^{(1)} = Z_{DUT} - Z_{PROSHORT} \qquad (2.6)$$

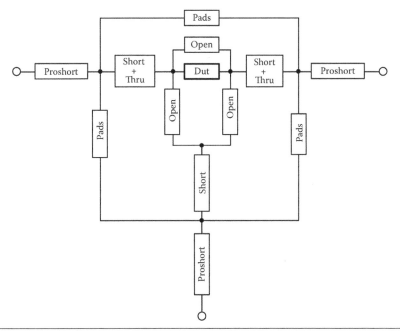

Figure 2.10 Impedance model of the five-step de-embedding technique. (From T. Biondi et al., 2004, "Characterization and Modeling of Sub-nH Integrated Inductances," *Proceedings of the IEEE Instrumentation and Measurement Technology Conference*, pp. 1998–2002.© 2004 IEEE. With permission.)

The second step removes the pad shunt parasitics by subtracting the Y-parameters of the pads from those of the DUT (and all remaining test structures), as reported in equation (2.7).

$$Y_{DUT}^{(2)} = Y_{DUT}^{(1)} - Y_{PADS} \qquad (2.7)$$

In the third and fourth steps the series parasitics of the metal interconnects are accounted for using measurements of the through and short structures, respectively. The third step consists in determining the equivalent series impedance of the through connection (Z_T), dividing it into two equal parts ($Z_{11} = Z_{22} = Z_T/2$), and subtracting it from the Z-parameters of the DUT (short and open), as reported in equations (2.8) and (2.9).

$$Z_{THRU} = \begin{bmatrix} Z_T/2 & 0 \\ 0 & Z_T/2 \end{bmatrix} \qquad (2.8)$$

$$Z_{DUT}^{(3)} = Z_{DUT}^{(2)} - Z_{THRU} \qquad (2.9)$$

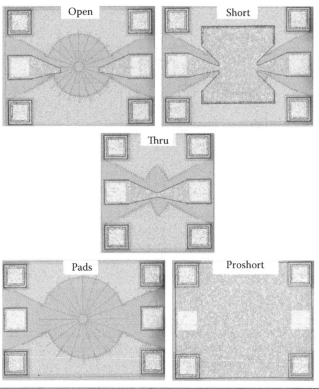

Figure 2.11 On-wafer test structures of the five-step de-embedding technique. (From T. Biondi et al., 2007, "Characterization and Modeling of Silicon Integrated Spiral Inductors for High-Frequency Applications," *Analog Integrated Circuits and Signal Processing*, vol. 51, pp. 89–100. © 2007 Springer. With permission.)

In the fourth step, the Z-parameters of the short are subtracted from those of the DUT (and open) according to equation (2.10).

$$Z_{\text{DUT}}^{(4)} = Z_{\text{DUT}}^{(3)} - Z_{\text{SHORT}} \qquad (2.10)$$

In the fifth step, the shunt parasitics of the metal interconnects are accounted for by subtracting the Y-parameters of the open structure from those of the DUT as reported in equation (2.11).

$$Y_{\text{DUT}}^{(5)} = Y_{\text{DUT}}^{(4)} - Y_{\text{OPEN}} \qquad (2.11)$$

This de-embedding procedure was compared with the most commonly used correction techniques reported in the literature, with the aim to investigate the achievable benefits in the measurements of sub-nH integrated inductors. Figures 2.12 through 2.14 report this comparison in terms of low-frequency inductance, peak quality factor,

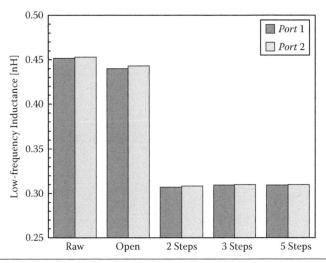

Figure 2.12 Comparison of de-embedding techniques on the low-frequency inductance of a 0.3-nH inductor with $n = 1.5$, $w = 14$ μm, $d_{in} = 50$ μm.

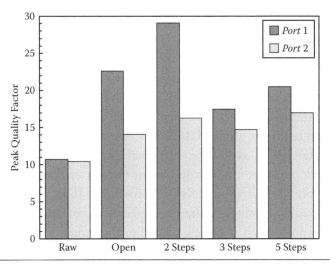

Figure 2.13 Comparison of de-embedding techniques on the peak quality factor of a 0.3-nH inductor with $n = 1.5$, $w = 14$ μm, $d_{in} = 50$ μm.

and peak quality factor frequency measured at both ports of the device (the underpass is located at port 2).

All de-embedding techniques that account for series parasitics are able to extract the proper value of low-frequency inductance from raw data. Indeed, the differences among the two-, three-, and five-step approaches are negligible at such low frequency. The only exception applies for the open correction method that predicts almost the same

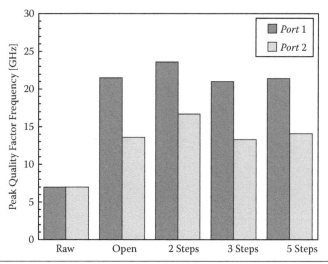

Figure 2.14 Comparison of de-embedding techniques on the peak quality factor frequency of a 0.3-nH inductor with $n = 1.5$, $w = 14$ µm, $d_{in} = 50$ µm.

inductance value as raw data. As expected, no appreciable differences can be observed between port 1 and port 2 measurements. The scenario changes completely when dealing with high-frequency measurements, as for the peak quality factor. The two-step method estimates the highest values of peak quality factor at both device ports; moreover, the measurement at port 1 estimates a peak quality factor of 29, which can hardly be attained in a silicon-based technology using only three metal layers, as will be explained later in this book. A more accurate description of series and shunt test pattern parasitics is provided by both the three- and five-step methods. In fact, both of them estimate realistic values of the peak quality factor and its frequency of occurrence, which is strictly related to the inductor self-resonance frequency. Discrepancies between the two methods are in the order of 15% for the peak quality factor and 5% for its frequency of occurrence. This difference must be attributed to the more accurate description of the series layout parasitics provided by the five-step impedance model described above and to the availability of two de-embedding test structures that are missing in the three-step method (i.e., pads and pro-short). It should be noted that the open correction method shows an apparent accuracy on the measurement of the peak quality factor and its frequency of occurrence. However, this is only due to a combination of errors on the real and imaginary parts of the measured impedance of the coil. Indeed, the actual accuracy of this method is clearly reported in Figure 2.12 where this method overestimates the low-frequency inductance (i.e., the imaginary part of the measured impedance) by more than 40%. A proportional error on the real part of such impedance might give the misleading idea that the quality factor measurement is roughly accurate.

2.4 Fabrication and Modeling: The State of the Art

Since 1990, when the feasibility of fabrication in silicon technology was first demonstrated (Nguyen and Meyer 1990), monolithic inductors have been widely investigated and many advances have been made in the fabrication and modeling on both silicon and gallium arsenide substrates. Two main directions have been followed during the years: improving the performance of the device by modifying the layout and/or the fabrication process to achieve either higher quality factor or lower area occupancy and developing numerical and/or circuit models to analyze the loss mechanisms that characterize the behavior of inductors built on a lossy substrate.

2.4.1 Fabrication Technology and Advanced Layout Techniques

As described earlier in this chapter, the loss mechanisms that influence the performance of monolithic inductors can be associated to the metal trace and substrate. Metal losses are caused by the finite resistivity of the conductors employed to fabricate the coil (most commonly aluminum or copper) and are responsible for the inductor series resistance. At higher frequencies skin and proximity effects take place, forcing the current to flow toward the outer surface of the conductor. This effect, known as *current crowding*, reduces the effective surface available for current conduction (with respect to the dc case) and increases the series resistance of the coil at higher frequencies.

The effects of the substrate on the performance of monolithic inductors may be much more critical than those caused by current crowding, especially at higher frequencies. Indeed, the electric coupling between the coil and the underlying substrate limits the maximum operating frequency of the device, causing self-resonance to occur, after which the capacitive behavior dominates over the inductive one. Although displacement currents electrically induced into the substrate also contribute to the overall losses of the device, the effect of magnetically induced currents is much more detrimental in standard silicon technologies. Indeed, the magnetic field generated by the coil induces an electric field into the substrate determining a current flow. Besides ohmic losses, whose magnitude depends on the resistivity of the substrate, these currents produce a magnetic field that tends to oppose to the impressed field, reducing the self-inductance of the coil. Moreover, they increase the effect of current crowding due to the proximity between the coil and substrate. This phenomenon is strongly reduced in monolithic inductors fabricated on gallium arsenide thanks to the semi-insulating properties of this substrate but represents the main cause of losses in silicon because even lowly doped substrates allow currents of significant magnitude to flow.

The most widely used technique to reduce metal losses basically consists in lowering the sheet resistance of the conductor by either increasing its conductivity and/or thickness. Because the most commonly employed conductors in silicon technologies are aluminum and copper, many researchers have successfully exploited gold to pursue the former approach, whereas the latter has been achieved by modifying the fabrication process (Ashby et al. 1996). Another technique exploits the evidence that current crowding is much more pronounced in the innermost turns of the coil; therefore, series losses can be reduced by designing hollow inductors (Craninckx and Steyaert 1997) or by tapering the conductor to reduce series losses and substrate coupling at the same time (Lopez-Villegas et al. 2000). Another solution consists in shunting two or more spirals, stacked one over the other on different metal layers, in order to obtain an equivalent conductor with increased thickness. This technique reduces the series resistance without additional fabrication costs; however, the use of lower metal layers increases both magnetic and electric coupling with the substrate (Soyuer et al. 1995). Despite substrate detrimental effects, this solution becomes necessary in technologies featuring very thin conductors (e.g., VLSI CMOS processes). A series (instead of shunt) connection between stacked coils can also be exploited to reduce series losses because inductance increases at a higher rate than resistance, resulting in higher quality factors (Merrill et al. 1995). The coils can be laterally or diagonally shifted to reduce the electric coupling between them.

Because substrate coupling is much more detrimental than current crowding in the metal layers, many efforts have been undertaken to reduce the losses deriving from its presence. Intuitively, the simplest way to reduce substrate coupling is removing it under the spiral, which can be accomplished in silicon technology through selective etching (Chang, Abidi, and Gaitan 1993). The spiral is now suspended over an air cushion, which is insulating enough to render current crowding the most limiting factor at the frequencies of interest. As an alternative solution to the costly and complicated process of substrate etching, inductors fabricated on insulating materials, such as sapphire, plastic, or glass, have been successfully proposed (Johnson et al. 1996; Dekker et al. 1997). Among the so-called unconventional technologies, micromachined solenoid-type inductors must be enumerated as the most successful attempt to steal discrete fabrication of its advantages and transfer them to the planar technology (Yoon et al. 1998).

Less expensive solutions have also been proposed that exploit pure silicon technologies. It was demonstrated that both magnetically induced and displacement currents can be effectively reduced employing a highly resistive silicon substrate (Park et al. 1997). Reverse-biasing the substrate junction has also been exploited to enhance isolation of the spiral, because lowly doped substrates generate deep depletion regions (Kim and O 1997). On the contrary, employing a polysilicon or metal pattern to electrically shield the spiral from the substrate and provide a low-impedance return path to ground proved successful to reduce substrate losses by hampering the flow of eddy currents and minimizing the voltage drop caused by displacement currents (Yue and Wong 1997). The main drawback of this solution is the reduced self-resonance frequency that derives from the higher capacitance between the spiral and the ground shield. In either highly or lowly doped substrates, increasing the physical separation between the spiral and substrate by thickening the silicon oxide helps to reduce both electric and magnetic coupling at the expense of increased fabrication costs.

2.4.2 Numerical and Circuit Modeling

Full three-dimensional electromagnetic simulation is, with no doubt, the most suited tool to accurately analyze the involved phenomena that give rise to energy loss in monolithic inductors and transformers.

Although many efforts have been undertaken to realize efficient algorithms to solve Maxwell's equations in three dimensions, commercially available simulators still require too much resources in terms of both memory and computation time. As a consequence, they can hardly be employed as an optimization tool and are commonly replaced by simplified numerical methods or equivalent circuit models.

One way to capture the distributed nature of monolithic inductors consists in dividing the length of the spiral into segments of transmission line subject to mutual magnetic coupling. Each segment can be modeled employing the simple per-unit-length equivalent circuit of a transmission line where the section of substrate underlying the segment represents the ground reference. This concept forms the basis upon which most distributed circuit models are developed. The elements that compose each segment are calculated using numerical techniques to solve Maxwell's equations in one or two dimensions. In Long and Copeland (1997) the self-inductance of each segment is calculated using the Greenhouse method, the series resistance increases with frequency only due to skin effect, a two-dimensional model is employed to calculate the substrate capacitance, and free-space Green functions are employed to account for substrate effects. Current crowding caused by the proximity effect is neglected because uniform current distribution is considered within the conductor and magnetically induced currents into the substrate are modeled indirectly.

The nonuniform current distribution that takes place within the coil is analyzed in Niknejad and Meyer (1998, 2001), exploiting the concept of partial element equivalent circuit, where the section of the conductor is divided into smaller parts where uniform current distribution is assumed. The total (frequency-dependent) inductance and resistance, subject to skin and proximity effects, result from the mutual magnetic coupling among all the parts in which the section has been divided. Moreover, a three-dimensional multilayer Green function is employed to analyze the magnetically and electrically induced current into the substrate. A simulation tool (ASITIC) was developed to solve the electromagnetic formulation of the problem and generate a lumped-element equivalent circuit model that approximates the inductor characteristics over a reasonably wide frequency range.

In spite of the accuracy and versatility offered by numerical techniques and related distributed networks, lumped equivalent circuits remain the most widely employed models among the RF IC designer community. Model components are calculated by means of physics-based or empirical closed-form expressions derived from the analytical solution of one- or two-dimensional electromagnetic problems under simplifying assumptions. In most of the models reported in the literature, expressions derived in previous papers are collected and applied (with minor modifications) to a given topology in order to capture the inductor behavior, whereas few model components are calculated using formulas developed for that specific case. The low-frequency inductance of the coil is commonly calculated by either closed-form expressions (Mohan et al. 1999) or applying the Greenhouse method (Greenhouse 1974). The rise of series resistance caused by skin and proximity effects is taken into account by adding the dc series resistance with two terms that depend on frequency according to square-root and square relationships, respectively, whose coefficients are obtained by least square fitting experimental data (Long and Danesh 2001). A physics-based expression for the series resistance of multiturn coils subject to current crowding is reported in Kuhn and Ibrahim (2001). Besides increasing the series resistance, current crowding is also responsible for a slight reduction of inductance at higher frequencies. Although the latter phenomenon can be safely neglected in most cases, a ladder RL network can be employed to take both effects into account (Cao et al. 2002). Most commonly, the substrate, underpass, and fringe capacitances of the model are treated as parallel-plate capacitances whose value is proportional to the area (and, in some cases, perimeter) of the opposite surfaces. More accurate expressions for the distributed capacitance of coils can be found in Jang, Excell, and Hejazi (1997).

The substrate impedance is commonly related to the electrical and geometrical characteristics of the ground path. Equivalent circuits different from the classical π-like topology, sometimes split into two or more sections, can be used to extend the accuracy of the model at higher frequencies and capture the electrical characteristics of a wider range of inductor geometries.

References

K. B. Ashby, I. A. Koullias, W. C. Finley, J. J. Bastek, and S. Moinian, 1996, "High Q Inductors for Wireless Applications in a Complementary Silicon Bipolar Process," *IEEE Journal of Solid-State Circuits*, vol. 31, pp. 4–9.

T. Biondi, A. Scuderi, E. Ragonese, and G. Palmisano, 2004, "Characterization and Modeling of Sub-nH Integrated Inductances," *Proceedings of the IEEE Instrumentation and Measurement Technology Conference*, pp. 1998–2002.

T. Biondi, A. Scuderi, E. Ragonese, and G. Palmisano, 2007, "Characterization and Modeling of Silicon Integrated Spiral Inductors for High-Frequency Applications," *Analog Integrated Circuits and Signal Processing*, vol. 51, pp. 89–100.

Y. Cao, R. A. Groves, N. D. Zamdmer, J.-O. Plouchart, R. A. Wachnik, X. Huang, T. J. King, and C. Hu, 2002, "Frequency-Independent Equivalent Circuit Model for On-Chip Spiral Inductors," *Proceedings of the IEEE Custom Integrated Circuit Conference*, pp. 217–220.

Cascade Microtech Inc., 1994, *A Guide to Better Vector Network Analyzer Calibrations for Probe-Tip Measurements*, Technical brief, Cascade Microtech, Inc., Beaverton, OR.

Cascade Microtech Inc., 1998, *On-Wafer Vector Network Analyzer Calibration and Measurements*, Technical brief, Cascade Microtech, Inc., Beaverton, OR.

J. Y.-C. Chang, A. A. Abidi, and M. Gaitan, 1993, "Large Suspended Inductors on Silicon and Their Use in a 2 mm CMOS RF Amplifier," *IEEE Electron Device Letters*, vol. 14, pp. 246–248.

H. Cho, and D. E. Burk, 1991, "A Three-Step Method for the De-Embedding of High-Frequency S-Parameters Measurements," *IEEE Transactions on Electron Devices*, vol. 38, pp. 1371–1375.

J. Craninckx, and M. Steyaert, 1997, "A 1.8-GHz Low-Phase-Noise CMOS VCO Using Optimized Hollow Spiral Inductors," *IEEE Journal of Solid-State Circuits*, vol. 32, pp. 736–744.

A. Davidson, E. Strid, and K. Jones, 1989, *Achieving Greater On-Wafer S-Parameter Accuracy with the LRM Calibration Technique*, Cascade Microtech Inc., Beaverton, OR.

R. Dekker, P. Baltus, M. Van Deurzen, W. v. d. Einden, H. Maas, and A. Wagemans, 1997, "An Ultra Low-Power RF Bipolar Technology on Glass," *International Electron Devices Meeting*, pp. 921–923.

A. Frasen, R. Gleason, and E. W. Strid, 1988, "GHz On-Silicon-Wafer Probing Calibration Methods," *Proceedings of the IEEE Bipolar/BiCMOS Circuits and Technology Meeting*, pp. 154–157.

M. Greenhouse, 1974, "Design of Planar Rectangular Microelectronics Inductors," *IEEE Transactions on Parts, Hybrids, and Packaging*, vol. 10, pp. 101–109.

Z. Jang, P. S. Excell, and Z. M. Hejazi, 1997, "Calculation of Distributed Capacitances of Spiral Resonators," *IEEE Transactions on Microwave Theory and Techniques*, vol. 45, pp. 139–142.

R. A. Johnson, C. E. Chang, P. M. Asbeck, M. E. Wood, G. A. Garcia, and I. Lagnado, 1996, "Comparison of Microwave Inductors Fabricated on Silicon-on-Sapphire and Bulk Silicon," *IEEE Microwave and Guided Wave Letters*, vol. 6, pp. 323–325.

K. Kim, and K. O, 1997, "Characteristics of an Integrated Spiral Inductor with an Underlying n-Well," *IEEE Transactions on Electron Devices*, vol. 44, pp. 1565–1567.

T. E. Kolding, 2000, "A Four-Step Method for De-Embedding Gigahertz On-Wafer CMOS Measurements," *IEEE Transactions on Electron Devices*, vol. 47, pp. 734–740.

M. C. A. M. Koolen, J. A. M. Geelen, and M. P. J. G. Versleijen, 1991, "An Improved De-Embedding Technique for On-Wafer High-Frequency Characterization," *Proceedings of the IEEE Bipolar/BiCMOS Circuits and Technology Meeting*, pp. 188–191.

W. B. Kuhn, and N. M. Ibrahim, 2001, "Analysis of Current Crowding Effects in Multiturn Spiral Inductors," *IEEE Transactions on Microwave Theory and Techniques*, vol. 49, pp. 31–38.

S. Lautzenhiser, A. Davidson, and K. Jones, 1990, "Improve Accuracy of On-Wafer Tests via LRM Calibration," *Microwaves & RF*, vol. 29, pp. 105–109.

J. R. Long, and M. A. Copeland, 1997, "The Modeling, Characterization, and Design of Monolithic Inductors for Silicon RF IC's," *IEEE Journal of Solid-State Circuits*, vol. 32, pp. 357–368.

J. R. Long, and M. Danesh, 2001, "A Uniform Compact Model for Planar RF/MMIC Interconnects, Inductors and Transformers," *Proceedings of the IEEE Bipolar/BiCMOS Circuits and Technology Meeting*, pp. 167–170.

J. M. Lopez-Villegas, J. Samitier, C. Cane, P. Losantos, and J. Bausells, 2000, "Improvement of the Quality Factor of RF Integrated Inductors by Layout Optimization," *IEEE Transactions on Microwave Theory and Techniques*, vol. 48, pp. 76–83.

A. J. Lord, 2000, *Comparing the Accuracy and Repeatability of On-Wafer Calibration Techniques to 110 GHz*, Cascade Microtech Inc., Beaverton, OR.

J. C. Maxwell, 1865, *A Dynamical Theory of the Electromagnetic Field*. Oxford University Press, Oxford, U.K.

J. C. Maxwell, 1873, *A Treatise on Electricity & Magnetism*. Philosophical Transection of the Royal Society of London, London, U.K.

R. B. Merrill, T. W. Lee, H. You, R. Rasmussen, and L. A. Moberly, 1995, "Optimization of High Q Integrated Inductors for Multi-Level Metal CMOS," *International Electron Devices Meeting*, pp. 983–986.

S. S. Mohan, M. del Mar Hershenson, S. P. Boyd, and T. H. Lee, 1999, "Simple Accurate Expressions for Planar Spiral Inductances," *IEEE Journal of Solid-State Circuits*, vol. 34, pp. 1419–1424.

N. M. Nguyen, and R. G. Meyer, 1990, "Si IC-Compatible Inductors and LC Passive Filters," *IEEE Journal of Solid-State Circuits*, vol. 27, pp. 1028–1031.

A. M. Niknejad, and R. G. Meyer, 1998, "Analysis, Design and Optimization of Spiral Inductors and Transformers for Si RF ICs," *IEEE Journal of Solid-State Circuits*, vol. 33, pp. 1470–1481.

A. M. Niknejad, and R. G. Meyer, 2001, "Analysis of Eddy-Current Losses over Conductive Substrates with Applications to Monolithic Inductors and Transformers," *IEEE Journal of Solid-State Circuits*, vol. 49, pp. 166–176.

M. Park, C. S. Kim, J. M. Park, H. K. Yu, and K. S. Nam, 1997, "High Q Microwave Inductors in CMOS Double-Metal Technology and Its Substrate Bias Effects for 2 GHz RF ICs Application," *International Electron Devices Meeting*, pp. 59–62.

M. Soyuer, J. N. Burghartz, K. A. Jenkins, S. Ponnapalli, J. F. Ewen, and W. E. Pence, 1995, "Multilevel Monolithic Inductors in Silicon Technology," *Electronics Letters*, vol. 31, pp. 359–360.

E. P. Vandamme, D. M. M.-P. Schreurs, and C. van Dinther, 2001, "Improved Three-Step De-Embedding Method to Accurately Account for the Influence of Pad Parasitics in Silicon On-Wafer RF Test-Structures," *IEEE Transactions on Electron Devices*, vol. 48, pp. 737–742.

P. J. van Wijnen, H. R. Claessen, and E. A. Wolsheimer, 1987, "A New Straightforward Calibration and Correction Procedure for 'On-Wafer' High-Frequency S-Parameter Measurements (45 MHz–18 GHz)," *Proceedings of the IEEE Bipolar/BiCMOS Circuits and Technology Meeting*, pp. 70–73.

J. Weng, 1995, "A Universal De-Embedding Procedure for the 'On-Wafer' GHz Probing," *IEEE Transactions on Electron Devices*, vol. 32, pp. 1703–1705.

J.-B. Yoon, B.-K. Kim, C.-H. Han, E. Yoon, K. Lee, and C.-K. Kim, 1998, "High-Performance Electroplated Solenoid-Type Integrated Inductor (SI²) for RF Applications Using Simple 3D Surface Micromachining Technology," *International Electron Devices Meeting*, pp. 544–547.

C. P. Yue, and S. S. Wong, 1997, "On-Chip Spiral Inductors with Patterned Ground Shields for Si-Based RF IC's," *IEEE Symposium on VLSI Circuits*, pp. 85–86.

3
MONOLITHIC INDUCTORS ON SILICON

Monolithic inductors and transformers are widely used to improve the performance of most RF ICs thanks to their low cost and ease of integration. They have wide application in power and noise matching for low-noise amplifier (LNA), resonant loads, degeneration, and linearity of amplifiers and mixers. Magnetic components basically provide finite AC impedance with low DC losses and, thanks to this property, they allow circuits to be biased at the supply voltage optimizing the voltage swing and/or maximizing the linearity of the component. Inductors and transformers are also largely used as resonant loads and matching networks in RF power amplifiers, whose efficiency is affected by passive losses. Magnetically coupled monolithic components have a fundamental role in the design of low-noise voltage-controlled oscillator (VCO). Integrated tanks are mandatory at gigahertz frequencies because external components are usually limited by spurious resonances due to parasitic capacitances and bonding wire inductances. The performance in terms of phase noise and oscillation amplitude with respect to current consumption depends on the Q of the LC resonator. Finally, RF filters also are built using monolithic components especially at high operating frequency.

However, the design of inductors in RF and mm-wave circuits is an issue of great concern. An accurate model taking into account all loss phenomena is a hard task and the only way to accurately predict high-frequency performance is to use electromagnetic (EM) simulators. Unfortunately, these tools are very time and resource consuming and do not meet the requirements of RF designers who need simple and accurate models for design and optimization.

In this chapter a treatment of monolithic inductors for RF IC in silicon technology is presented, providing the reader with both design and lumped modeling issues. The chapter is organized as follows.

A brief overview on modern silicon-based technologies is provided in Section 3.1 with a description of the process used in this work. Inductor substrate analysis based on experimental comparison is provided in Section 3.2, highlighting the benefits of patterned ground shields in term of component performance and modeling. Moreover, in order to carry out more profitable discussion, a large set of inductors was integrated in a silicon bipolar technology and presented in Section 3.3. An analysis of inductance equations reported in literature is carried out in Section 3.4, where expressions for inductances in the range of 0.1–10 nH are compared to measures. A physics-based lumped scalable model for monolithic inductors is validated by experimental comparison in Section 3.5.

3.1   Overview on Silicon Technology

3.1.1   Silicon Technologies

At the time of writing, state-of-the-art technologies feature SiGe heterojunction bipolar transistor (HBT) approaching 300 GHz for both f_T and f_{max}, as reported in Avenier et al. (2009) and Knapp et al. (2007). The aim of such technologies is to approach mm-wave field, where applications such as 60-GHz WLAN, 77-GHz radar sensors, and 100 Gb/s optical communication require active devices with high cutoff frequency and low-noise performance, supported by low-loss back-end for high-Q passive devices. In Avenier et al. (2009), a BiCMOS technology offers a complete platform for modern complex transceivers, which require low-noise signal treatment, watt-level power delivery, and digital processing. Indeed, the technology features a very high-speed performance SiGe HBT with a 240/270 GHz f_T/f_{max} and minimum noise figure below 4 dB even at 100 GHz. A medium-voltage HBT is also embedded for power delivery and a 0.13-μm CMOS platform is suitable for the integration of thousands of gates. The back-end is optimized to integrate high-Q passive devices, which can exploit two 3-μm-thick copper metal layers fabricated over two 3-μm-thick intermetal dielectrics. As a consequence, losses due to series resistance and parallel capacitance with substrate are minimized.

Nanometer pure CMOS technologies take advantage from scaling the channel length down to 28 nm, achieving f_T and f_{max} well above 300 GHz (Li et al. 2007). Achieved low-noise performance promotes very short-channel MOS as a promising technology for high-frequency applications. Actually, the main drawback of CMOS technologies is the back-end, usually composed of many copper metal layers (even more than 10), but most of them are very thin, with thickness lower than 0.5 μm. This aspect limits the quality of passive devices at high operating frequencies. Today, this drawback is overcome in CMOS technologies oriented for RF applications, where the upper metal levels are typically 3 μm thick. In other CMOS processes, the Q of passive devices is improved by stacking more metal levels to build the device. In this case series losses are reduced at the cost of a self-resonance frequency reduction due to a higher substrate parasitics.

3.1.2   Fabrication Technology

Design and modeling of inductive devices require a full knowledge of the process back-end and thickness of metals and intermetal dielectric layers. For this reason, hereafter, a short description of the fabrication technology used in this work is provided. A 46-GHz f_T double poly 0.8-μm self-aligned emitter silicon bipolar process, whose simplified cross section is shown in Figure 3.1, is used (Ragonese et al., 2004). This low-cost technology requires only 19 mask steps and

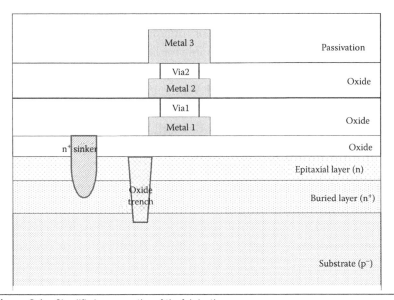

Figure 3.1   Simplified cross section of the fabrication process.

provides a p-channel MOSFET as a complementary device. It features oxide trench isolation, three AlSiCu metal layers, polyresistors, metal-insulator-metal (MIM) capacitors, and junction diode varactors. The three metal layers allowed by the process are profitably used to produce inductive devices, such as inductors and transformers. The highest metal level (Metal 3) is used to fabricate the coils, whereas the more resistive layers (Metal 2 and Metal 1) are used for contact paths and ground planes surrounding circuits and devices. The third, second, and first AlSiCu metal layers have thickness equal to 3, 1, and 0.5 µm, respectively. The minimum Metal 3 spacing (s) allowed by the technology is 3 µm. The process allows n$^+$ sinkers to contact the buried layer to the Metal 1. Two optional mask steps provide a selective buried layer and p$^+$ substrate contacts, as well.

3.2   Substrate Optimization for Inductor on Silicon

In VLSI standard technologies, inductor performance, such as Q and self-resonance frequency (f_{SR}) is mainly limited by electromagnetic phenomena, taking place within the substrate layers underlying the spiral, as discussed in Chapter 2. Considerable efforts were provided to reduce substrate losses, but results have been often disappointing in comparison with the additional costs required. Higher performance has been achieved by exploiting unconventional technologies—that is, silicon on insulator (SOI) or microelectromechanical systems (MEMS)—but these are not suitable for large volume production due to their high fabrication costs. Solutions based on substrate shielding are generally used (Yue and Wong 1998) in silicon technology, to avoid detrimental effects of eddy currents in the conductive substrate. Substrate shielding improves both the lumped modeling and EM simulations of passive devices, allowing an accurate estimation of the performance up to the f_{SR}.

   In order to understand the impact of the substrate on inductors and transformers, several components were fabricated on different vertical structures and a comparative analysis was carried out drawing benefits and drawbacks (Ragonese et al. 2004). A patterned ground shield (PGS) built on a buried layer allows inductive device performance to be maximized. Furthermore, both modeling and accurate EM simulations can be easily carried out.

3.2.1   Comparative Analysis

The performance degradation due to electromagnetic coupling between coils and the silicon substrate can be strongly reduced by using proper arrangements for conductive layers underlying the spiral. In particular, it is mandatory to avoid the detrimental effects due to the highly doped n^+ buried layer. Indeed, magnetically induced currents on this layer reduce both the inductance and Q, whereas capacitive coupling determines the f_{SR}.

In order to overcome these drawbacks, two different strategies are analyzed. The first approach aims to replace silicon conductive layers with a honeycomb pattern of oxide trench that hampers planar conduction loops. Following the same idea, the buried layer can be selectively removed under the spiral by exploiting additional mask steps of the process. These approaches also reduce electrical coupling thus improving the f_{SR}.

An alternative solution takes advantage of the highly conductive layers below the coil in order to build a ground shield. An oxide trench pattern is used to cut the highly doped n^+ buried layer, under the spiral orthogonally to current loops. This proposed structure achieves better performance than Metal 1 PGS, because it also preserves f_{SR}. Vertical structures described above were tested and compared on both inductors and transformers (the same analysis for transformer is addressed in Chapter 4).

An octagonal spiral (w = 6 μm, n = 3.5, s = 3 μm, and d_{in} = 110 μm) was fabricated on honeycomb oxide trenched buried layer, Metal 1 PGS, and buried layer PGS. Die micrographs of each inductor are shown in Figure 3.2. De-embedding structures were also fabricated to eliminate layout test pattern effects. The investigated structures are compared in terms of Q and f_{SR} in Figure 3.3.

Experimental results revealed a 32% peak improvement on Q for the buried layer PGS with respect to the oxide trench structure. Moreover, in spite of a lower f_{SR}, the Q peak appeared at a higher frequency, proving that the effective operating band had not been reduced. Besides the expected f_{SR} increase, inductors on buried layer PGS reveal a 20% Q enhancement respect to Metal 1 PGS. Finally, based on experimental data no further improvement is achieved by removing the buried layer with respect to trench oxide structure.

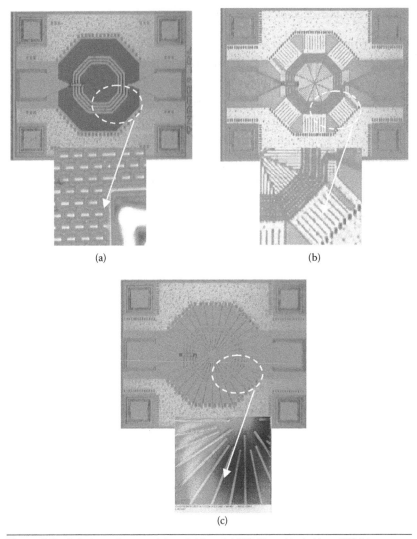

(a) (b)

(c)

Figure 3.2 Inductor on honeycomb trench (a), metal PGS (b), and buried layer PGS (c).

3.2.2 Benefits of Buried Layer PGS

Buried layer PGS for inductive devices achieves better performance and allows some design drawbacks to be overcome as well. In particular, the proposed structure naturally solves cross-talk problems and simplifies substrate modeling. Losses and cross-talk are both reduced because buried layer PGS avoids magnetically induced currents and provides a low-resistance return path to RF ground. To enhance the shielding effect each sector of the PGS must be contacted to the

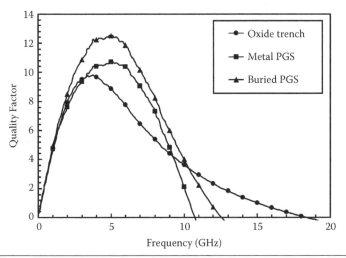

Figure 3.3 Quality factor comparisons. (From E. Ragonese et al., 2004, "Experimental Comparison of Substrate Structures for Inductors and Transformers," *IEEE MELECON Symposium Technical Digest,* pp. 143–146. © 2004 IEEE. With permission.)

ground metal plane in close proximity to the coil. Unfortunately, the ground metal plane greatly influences inductor performance because it affects both the f_{SR} and the Q. A 50-µm trade-off distance between the ground plane contacts and the coil is advised by measurements of several inductors with different ground plane distances. As an example, Figure 3.4 shows typical Q curves at two operating frequencies as a function of the ground plane distance for a circular inductor (w = 10 µm, n = 3.5, s = 3 µm, and d_{in} = 100 µm).

As explained before, substrate modeling for inductive components is a hard task, but PGS provides a well-defined RF ground reference, which is also simple to be modeled. Thanks to this benefit, in the next sections lumped modeling can be easily developed for devices with inductance in the range 0.1–10 nH.

3.3   Fabricated Inductors

In order to provide the reader with more profitable discussions regarding modeling and design, a wide set of single-layer circular inductors with different geometrical parameters, both for single-ended and differential circuits, was integrated and characterized. The inductors were fabricated in silicon technology by using the third metal layer (Metal 3) for the spiral, the second one (Metal 2) for the underpass,

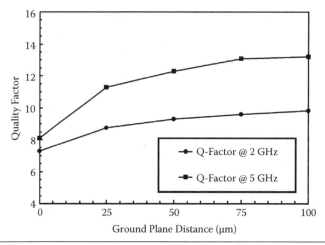

Figure 3.4 Quality factor at 2 and 5 GHz as function of ground plane distance. (From E. Ragonese et al., 2004, "Experimental Comparison of Substrate Structures for Inductors and Transformers," *IEEE MELECON Symposium Technical Digest*, pp. 143–146. © 2004 IEEE. With permission.)

and the first one (Metal 1) for the ground plane. According to the consideration proposed in Section 3.2, a radial pattern of oxide trench was adopted to break induced current loops within the buried layer, shielding the spiral from the underlying substrate. Buried layer contacts were placed at the edge of the ground plane at a distance of 50 μm from the spiral. Metal spacing was set to 4 μm.

3.3.1 Integrated Inductors for Single-Ended Applications

Integrated inductors for single-ended excitation have width from 6 to 20 μm and inner diameter from 50 to 150 μm as detailed in Table 3.1, where they are classified into 15 types. Each type has turn number of 2.5, 3.5, 4.5, and 5.5. The only exception applies for 18- and 20-μm-wide inductors that are limited to 3.5 and 2.5 turns, respectively, because f_{SR} of larger inductors is too low for RF applications. The resulting inductances vary from 0.3 to 8.6 nH covering most of the values employed in the design of silicon RF ICs.

In the following, inductors are labeled by type and turn number (e.g., G2.5). For each inductor, de-embedding structures were fabricated and measured up to 50 GHz to eliminate layout test pattern parasitics. A microphotograph of an integrated inductor is shown in Figure 3.5.

Table 3.1 Classification of Single-Ended Inductors

TYPE	w (μm)	d_{in} (μm)	TYPE	w (μm)	d_{in} (μm)	TYPE	w (μm)	d_{in} (μm)
A	6	50	F	10	150	M	18	100
B	6	100	G	14	50	N	18	150
C	6	150	H	14	100	O	20	50
D	10	50	I	14	150	P	20	100
E	10	100	L	18	50	Q	20	150

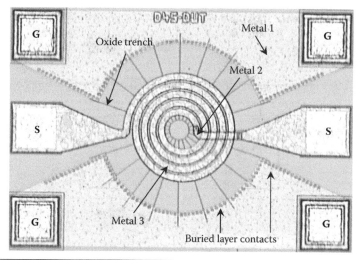

Figure 3.5 Micrograph of integrated inductor for single-ended topologies ($w = 10$ μm, $d_{in} = 50$ μm, $n = 4.5$).

3.3.2 Symmetric Integrated Inductors

Because RF ICs are widely based on differential topology to exploit the common node benefits, symmetrical inductors can be profitably used to drive and/or load differential pairs. Instead of two asymmetric separate spirals, a symmetric coil occupies lower area providing the full electrical symmetry.

An inductor excited by a differential source presents inherent benefits in terms of Q and f_{SR} with respect to the case of a single-ended excitation. This concept can be easily explained by modeling the inductor with a simple π-network, as shown in Figure 3.6(a), where $R(f)$ represent the series losses occurring in the metal, and Z_S models the parasitic elements of substrate. Z_S is composed of a capacitor, C_{OX}, which models the capacitive parasitic between metal layers and substrate, and an RC network, which represents the substrate. In

Figure 3.6(b) and Figure 3.6(c) the π-model is excited in single- and differential-ended configuration, respectively. The inductor excited by differential source compared to the single-ended experiments lower parasitic capacitance, $2Z_S$ instead of Z_S, providing higher Q and f_{SR} (Danesh, and Long 2002). Because f_{SR} depends on substrate capacitive parasitic, inductors excited differentially exhibit a frequency range of usage roughly doubled with respect to single-ended excited ones. Moreover, at working frequencies where substrate losses predominate, differentially driven devices present a Q enhancement. At low

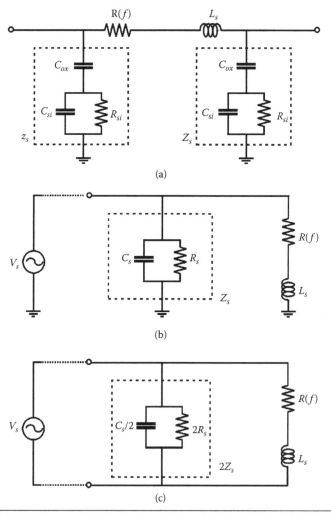

(a)

(b)

(c)

Figure 3.6 Inductor π-model (a) in single-ended (b) and differential driven (c) configuration.

frequencies, where the Q is affected by metal losses, no additional improvements are provided by differential excitation.

A wide set of symmetric inductors was fabricated ranging the turn number from 1 to 5, width from 6 to 14 µm, and inner diameter from 50 to 150 µm as detailed in Table 3.2, where they are classified into nine types. In the following, inductors will be labeled by type and turn number (e.g., Ad3). For each inductor, de-embedding structures were fabricated and measured up to 50 GHz to eliminate layout test pattern parasitic. A microphotograph of symmetric integrated inductor is shown in Figure 3.7.

Table 3.2 Classification of Symmetric Inductors

TYPE	w (µm)	d_{in} (µm)	TYPE	w (µm)	d_{in} (µm)
Ad	6	50	Gd	14	50
Bd	6	100	Hd	14	100
Cd	6	150	Id	14	150
Dd	10	50			
Ed	10	100			
Fd	10	150			

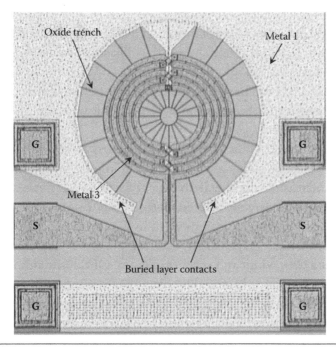

Figure 3.7 Micrograph of symmetric inductor ($w = 10$ µm, $d_{in} = 100$ µm, $n = 5$).

3.4   Inductance Calculation

A proper estimation of the inductance is the first step in the design of monolithic inductors. Exact calculation can be carried out by solving Maxwell equations. Consequently, full-wave 3D simulators can provide accurate prediction and thanks to the computational power of actual machines they can be considered not so time consuming to solve EM problems. In spite of this consideration, in the design flow of RF and mm-wave circuits, EM simulators are usually considered a tool of verification and not a tool of optimization. In the case of inductors, simple lumped scalable models are largely employed in design optimization. Their soundness is strictly related to the availability of accurate closed-form expressions for all the lumped elements and especially for the low-frequency inductance, whose estimation affects both the f_{SR} and Q calculations.

Several papers concerning the calculation of inductance have been published and meaningful results have been obtained. Unfortunately, closed-form equations still provide poor prediction in the calculation of inductance for low-value inductors ($L < 1$ nH, namely sub-nH inductors in the following) and thick-metal coils; that is, inductors featuring high thickness-to-width ratio.

Sub-nH inductances are of major interest for high-frequency applications (Ku, K, and Ka band), where the inductance values required by integrated circuits fall into the sub-nH range. Moreover, inductances larger than 1 nH can hardly be used at higher frequencies in standard silicon technologies because the tight coupling with the underlying substrate limits the f_{SR} and Q.

Because several efforts have been addressed to reduce the metal sheet resistance (Coolbaugh et al. 2002) and raise the Q at lower frequency, increasing the metal thickness was profitably investigated. Technological research is focused on the feasibility of very thick metals on silicon process as demonstrated in Figure 3.8, where the scanning electron micrograph (SEM) cross section of a 15-μm-thick metal is achieved with width and spacing of 6 μm. Obviously, although the fabrication is obtained, several problems concerning minimum spacing and width arise and foster further studies on the achievable performance of thick-metal inductors. Despite several experimental results on this topic that have been reported, few works faced the impact of

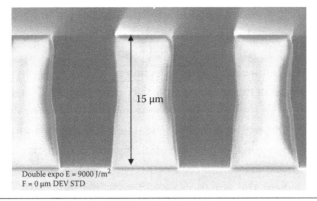

15 µm

Double expo E = 9000 J/m²
F = 0 µm DEV STD

Figure 3.8   SEM of 15-µm-thick metal.

metal thickening on the design, optimization, and modeling of spiral inductors (Choi and Yoon 2004; Scuderi et al. 2005). Moreover, much interest was focused on the increase of Q, overlooking the reduction of inductance. Although this phenomenon is taken into account in many expressions reported in the literature (Mohan 1999; Yue and Wong 2000; Jenei, Nauwelaers, and Decoutere 2002), their validation was limited to medium-thickness inductors with thickness-to-width ratios below 0.25. Therefore, commonly used inductance formulas produce unacceptable errors in state-of-the-art inductors where the metal thickness may even exceed the metal width.

In this section a brief review of inductance equations is provided, where maximum errors and inherent limitations of each equation are presented suggesting the equation based on current sheet approximation (Mohan 1999) as the most useful one. Moreover, an extension of the current sheet equation for sub-nH inductances is proposed to correct the unacceptable errors in the estimation of the inductance of low value asymmetric coils. Finally, a closed-form expression for low-frequency inductance calculation of thick-metal spiral inductors is proposed and validated in a wide range of geometrical parameters and thickness-to-width ratios.

3.4.1   State-of-the-Art Inductance Equations

Some analytical approaches for the inductance calculation are based on the Greenhouse method (Greenhouse 1974), which is accurate enough but not suitable for simple lumped modeling. To overcome this drawback, several closed-form expressions were published in last few years. They can be roughly divided into two different categories; that is, physics-based and monomial formulas.

In Jenei, Nauwelaers, and Decoutere (2002), a physics-based closed-form inductance expression was proposed and based on the decomposition of the coil into segments and calculation of the average interaction. The expression was calculated for square inductors and extended to polygonal cases. The accuracy was demonstrated by comparison with experimental data of polygonal inductors in the range 2–25 nH, where the equation provides estimations with the maximum error of 8%. Unfortunately, Jenei's equation is not so handy for designers and its accuracy decreases for small value inductors.

In Mohan et al. (1999), the current sheet approach provides simple, accurate expressions for the calculation of self- and mutual inductances of a variety of geometries. Approximating the sides of a spiral by symmetrical current sheet of equivalent current density, the authors avoid more complicated expressions based on summation methods. For example, the square spiral inductor can be approximated by using four identical current sheets. Using symmetry and the fact that sheets with orthogonal current have zero mutual inductance, the inductance can be calculated evaluating the self-inductance of one sheet and the mutual inductance between opposite current sheets. These self- and mutual inductances are evaluated using the concepts of geometric mean distance (GMD), arithmetic mean distance (AMD), and arithmetic mean square distance (AMSD). These calculations were generalized for polygonal geometries up to circular inductors. The obtained equation is reported in (3.1).

$$L_{orig} = \frac{\mu}{2} \cdot c_1 \cdot n^2 \cdot d_{avg} \cdot f(\rho) \qquad (3.1)$$

where

$$f(\rho) = \ln\left(\frac{c_2}{\rho}\right) + c_3\rho + c_4\rho^2 \qquad (3.2)$$

n is the number of turns, $d_{avg} = 0.5 \cdot (d_{in} + d_{out})$ is the average diameter (because d_{in} and d_{out} are the inner and outer diameters, respectively), $\rho = (d_{in} - d_{out})/(d_{in} + d_{out})$ is the fill factor, and the c_is are layout dependent parameters summarized in Table 3.3.

Table 3.3 Coefficients for Current Sheet Expression

SHAPE	c_1	c_2	c_3	c_4
Circular	1	2.46	0	0.2
Octagonal	1.07	2.29	0	0.19

Moreover, a monomial equation is proposed in Mohan (1999), obtained by least square fitting techniques using data extracted from extensive EM simulations carried out with ASITIC. Both equations were compared with measured data of 2.5–34 nH inductances and simulated coils in the range 0.5–100 nH. As demonstrated in Mohan et al. (1999), monomial and current sheet equations provide satisfactory results with median error about 4% and maximum error about 20%. However, errors higher than 10% are reported in few cases, which Mohan attributed to measurement inaccuracies. However, although the monomial equation provides good accuracy, its application is limited to polygonal geometry. Its extension to circular geometry implies the coefficient calculation from an extensive set of simulations. Moreover, the formulation of a monomial formula is more useful in optimization of inductors via geometric programming.

Current sheet equation is the most meaningful for inductance calculation because it was carried out from physics formulation of the problem. Unfortunately, to obtain a simple expression, some approximations, which limit the accuracy, are needed as explained above. The utmost hypothesis is the geometric symmetry; as a consequence, the inductance of symmetric coils is accurately predicted up to sub nH values, whereas errors as high as 19% were obtained for asymmetric low-value inductors (<1 nH). This inaccuracy can be ascribed to the high dissymmetry exhibited by the geometry of 1.5-turn inductors. Because the inductor values result in sub-nH range, the symmetry assumption supposed by the current sheet method results in excessive errors for such devices. A modified current sheet expression is introduced in Section 3.4.2 to overcome this problem in asymmetric coils, still maintaining simple formulation.

To confirm the soundness of the current sheet equation when symmetry is met, the error distribution, calculated between (3.1) and measured symmetric inductors, is presented in Figure 3.9. The error distribution curve provides useful information on the soundness of the expression. Relative error is reported on the x-axis, and the fraction of

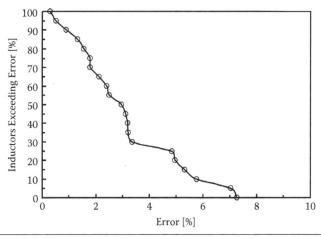

Figure 3.9   Error distribution of current sheet equation applied to symmetric coils.

inductors exceeding an error percentage level is reported on the y-axis. Such definition indicates more accuracy when curves are closer to the y-axis. Moreover, the errors corresponding to y level of 50 and 0% are the median and the maximum errors, respectively. In Figure 3.9, median and maximum errors of 3 and 7.2%, respectively, confirm the soundness of current sheet expression in the case of symmetric coils.

3.4.2   *Modified Current Sheet Expression for Sub-nH Inductors*

In this section the inherent limitations of the inductance equation for low-value asymmetric inductors are addressed. Fabricated devices result in inductances ranging from 0.3 to 8.6 nH, covering most of the values used in RF design. Self-resonance frequencies higher than about 20 GHz can only be achieved by inductances below 1 nH. Because it was observed experimentally that the maximum Q of the investigated inductors appears at approximately half the f_{SR}, inductances larger than 1 nH can hardly be used for high-frequency applications (Ku, K, and Ka band) in standard silicon technologies. This further demonstrates the need for characterization and modeling of sub-nH integrated inductors that were previously highlighted.

Due to the (weak) magnetic coupling with the ground plane and substrate, and due to the presence of the underpass, the inductance extracted from measurements differs from the value that competes to the spiral alone. Indeed, this value can only be computed through

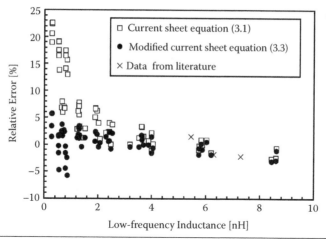

Figure 3.10 Relative errors of (3.1) and (3.3) with respect to EM simulations as a function of the low-frequency inductance. (Modified from T. Biondi et al., 2005, *IEEE Microwave and Wireless Components Letters*, 15: 922–924, © 2005 IEEE. With permission.)

EM simulations because the above parasitic cannot be avoided in the measurement setup. Simulations of open-air spirals (i.e., separated by a 5-mm air cushion from the underlying ground plane) are thus employed to validate the accuracy of the current sheet equations for circular spirals described in Section 3.4.1. In Biondi et al. (2005) a modified expression of equation (3.1) is reported as in (3.3).

$$L_{mod} = \frac{\mu}{2} \cdot \left(\frac{c_0}{n} + c_1 \cdot n^2 \right) \cdot d_{avg} \cdot f(\rho) \qquad (3.3)$$

In order to overcome the poor prediction for a low-value inductor, the additional fitting parameter c_0 is introduced to modify the dependence on the turn number, which is the geometrical parameter most affecting the inductance.

Relative errors of (3.1) and (3.3) with respect to EM simulations are shown in Figure 3.10 as a function of the low-frequency inductance. Both expressions provide few errors for inductances larger than 1 nH, but (3.1) becomes less accurate for smaller values with errors as large as 23%. On the other hand, choosing the value of c_0 (=0.73) in order to fit EM simulations provides maximum errors lower than 6% even for inductances as small as 0.3 nH, proving the effectiveness of the c_0 term for n between 1.5 and 5.5. The few experimental data concerning pseudocircular inductances (i.e., with 12 sides) taken from

Mohan et al. (1999) are also reported in Figure 3.10, confirming the trend of measurements.

Equation (3.3) can be easily extended to polygonal geometries. Indeed, in spite of their lower quality, polygonal inductors are still largely reported in the literature because they are easier to model using physics-based closed-form expressions. As an example, the validity of (3.3) was extended to octagonal inductors using the values of c_is reported in Table 3.3 and choosing a proper c_0 value (=0.57) in order to fit two-dimensional ½ EM simulations of a large set of sub-nH octagonal inductors. A comparison between the proposed expression and the state of the art is reported in Figure 3.11, where the error distribution (calculated with respect to EM simulations) of the expressions taken from Jenei, Nauwelaers, and Decoutere (2002) and Mohan et al. (1999) is compared with (3.3) in the range 0.3–1 nH. The worst cases occur for the modified Wheeler and current sheet expressions that reveal errors larger than 12% for more than 70% of the considered inductors, which is unacceptable for circuit design and optimization. The errors of the monomial (Mohan 1999) and physics-based (Jenei, Nauwelaers, and Decoutere 2002) equations do not exceed 10% with a median error (i.e., the error corresponding to 50% of inductors) as large as 6 and 4%, respectively. The most accurate prediction is obtained by the modified current sheet expression, which provides median and

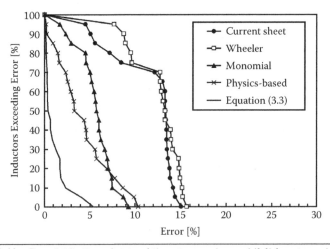

Figure 3.11 Error distributions of state-of-the-art expressions and (3.3) for octagonal spirals in the range 0.3–1 nH. (Modified from T. Biondi et al., 2005, *IEEE Microwave and Wireless Components Letters,* 15: 922–924, © 2005 IEEE. With permission.)

maximum errors lower than 1 and 6%, respectively. Again, the c_0 term improves the accuracy of the original expression in the sub-nH range, which confirms the soundness of the proposed approach.

It is worth noting that the c_0 term in (3.3) does not have a physical justification and only reflects the less-than-quadratic dependence on the turn number observed from experimental measurements at low inductance values. Although it (wrongly) provides infinitely large inductance values for turn number approaching zero, its decreasing monotonic behavior was verified for values of n down to 1. Moreover, the proposed expression can be employed for inductances as low as 0.1 nH.

3.4.3 Inductance Modeling of Thick-Metal Inductors

To investigate the effect of metal thickness on the low-frequency inductance of circular spirals, EM simulations (two-dimensional ½ EM simulator) have been used. To minimize the effect of substrate on simulated data, the coil inductance has been calculated from EM simulations of open-air spirals. The investigated structures have turn number from 1.5 to 5.5, inner diameter from 50 to 150 μm, and width from 6 to 20 μm. The geometrical parameters of all analyzed structures are detailed in Table 3.4. For each structure, the metal thickness t was varied from 3 to 15 μm. This results in thickness-to-width ratios from 0.15 to 2.5, which thoroughly exceeds the range commonly employed in the design of RF ICs.

By applying the current sheet approximation, the inductance of spirals with finite thickness conductor can be expressed by (3.4).

Table 3.4 Layout Parameters of Investigated Thick-Metal Inductors

TYPE	n	w (μm)	d_{in} (μm)
A_T	5.5	6	50
B_T	4.5	6	100
C_T	3.5	10	100
D_T	3.5	14	50
E_T	1.5	14	150
F_T	4.5	14	150
G_T	1.5	18	100
H_T	2.5	20	150

$$L = L_0 - \alpha \cdot \frac{\mu \cdot n^2 \cdot d_{avg}}{2} \cdot \frac{1}{n} \cdot \ln\left(1 + \frac{t}{w}\right) \qquad (3.4)$$

where L_0 is the inductance value that competes to a spiral with zero thickness, μ is the magnetic permeability of air, and d_{avg} is the average diameter computed as $(d_{in} + d_{out})/2$. The correction term α, equal to 1 in the original formulation of (3.4), has been introduced to provide more accurate inductance calculations for spirals with large values of n and t/w. Figure 3.12 shows the relative inductance decrease estimated by EM simulator, calculated as $100(1 - L/L_0)$, of spirals with different geometrical layout parameters as a function of the thickness-to-width ratio. In Figure 3.12, simulated data are also compared to the original (3.4) with $\alpha = 1$ and (3.4) with corrected α. It can be observed that the original expression underestimates the inductance reduction due to the increased metal thickness in all considered cases; moreover, the discrepancies between EM simulations and calculations become larger as the turn number increases. As an example, relative errors increase from 16% to more than 30% as the turn number rises from 1.5 to 5.5 even for thickness-to-width ratios smaller than 1. The formulation of (3.4) derives from the consideration that increasing the metal thickness only influences the self-inductance of the coil (proportional to n) while leaving unchanged the mutual inductance contributions

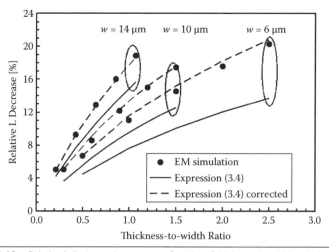

Figure 3.12   Relative inductance decrease as a function of thickness-to-width ratio for different widths.

(proportional to $n^2 - n$). However, questions might be raised on the validity of these assumptions in light of the above results. To improve the accuracy of (3.4), especially for large values of n, the correction term α was introduced. By comparing analytical calculations with EM simulations of spirals with different geometrical parameters it emerged that the dependence of (3.4) on the turn number has to be slightly modified in order to take into proper account the influence of the metal thickness on the mutual inductance contributions. This can be accomplished using an expression of the form $a \cdot n^b$, with a and b as fitting parameters, which allows accurate inductance modeling while still maintaining the physics-based nature of (3.4). The resulting expression of α is reported in (3.5).

$$\alpha = 1.13 \cdot n^{0.17} \tag{3.5}$$

Figure 3.12 demonstrates that the introduction of α in (3.4) substantially reduces calculation errors with respect to the original expression. Indeed, the corrected formula is in close agreement with EM simulations, providing errors smaller than 5% for turn number up to 5.5 and thickness-to-width ratio up to 2.5.

3.5   Modeling of Monolithic Inductors

The importance of inductor modeling has been discussed in several papers published in recent years. State-of-the-art modeling of silicon spiral inductors follows two different approaches: simple lumped π-equivalent networks (Yue and Wong 2000; Melendy et al. 2002) or distributed circuits (Long and Copeland 1997; Kythakyapuzha and Kuhn 2001). The former approach allows only low-frequency estimation of performance parameters and cannot be used over a wide range of layout geometries. The latter takes into account high-frequency effects and layout parameters as well, but it is quite difficult to manage for RF designers.

A frequency-dependent series resistance formula is commonly employed to take into account metal losses at higher frequencies (Kythakyapuzha and Kuhn 2001). Numeric methods, fitting parameters, or complex equations are implemented to ensure good agreement with experimental data over a wide range of geometries. Although inductors can be employed with either a grounded terminal (i.e., in emitter degenerations and resonant loads) or as a two-port device (i.e., in matching networks), for most of the models reported in the literature only one-port behavior has been validated by comparison with Y-parameter measurements.

In this section, a simple lumped scalable model for spiral inductors in silicon bipolar technology was developed and validated by comparison with experimental measurements over a wide range of geometrical layout parameters (Scuderi et al. 2004). The proposed model is based on a new topology, whose components are calculated by means of physics-based equations related to technological and geometrical parameters. An empirical equation for the series resistance was formulated to model the involved loss phenomena within the spiral. The model is scalable over a wide range of inductor geometries. Moreover, comparisons with both one- and two-port measured performance parameters revealed the same degree of accuracy up to frequencies well above self-resonance.

3.5.1   Lumped Scalable Model

The model shown in Figure 3.13 is proposed to overcome the main drawbacks of the classical π-model. Though simple π-like topologies are able to approximate inductance and quality factor, calculated as in (3.6) and (3.7), respectively, they do not properly take into account two-port behavior:

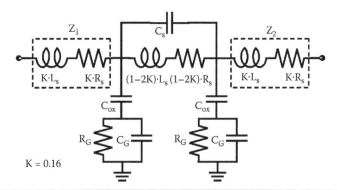

Figure 3.13   Proposed model for spiral inductors on PGS. (From A. Scuderi et al., 2004, "A Lumped Scalable Model for Silicon Integrated Spiral Inductors," *IEEE Transactions on Circuits and Systems Part I*, vol. 51, pp. 1203–1209. © 2004 IEEE. With permission.)

$$L = \frac{\operatorname{Im}\left(1/Y_{11}\right)}{\omega} \tag{3.6}$$

$$Q = -\frac{\operatorname{Im}\left(Y_{11}\right)}{\operatorname{Re}\left(Y_{11}\right)} \tag{3.7}$$

Adding Z_1 and Z_2 impedances ($Z_1 = Z_2$) allowed the π-model to be split, with K as the splitting factor. It thereby more closely approximates a distributed network while still maintaining the advantages of simplicity. The benefits of the proposed model with respect to the classical topology are demonstrated in Figure 3.14, where the measured and simulated S-parameters of the E3.5 inductor are plotted. It is clear that the π-model ($K = 0$) does not allow correct estimation of S_{11}, which provides the f_{SR}, and S_{12}, which is of utmost importance in modeling the two-port behavior. On the other hand, simulations of the proposed model with different values of the splitting factor showed that, even if a value of 0.1 is needed to provide a good agreement with measured Y-parameters, the best trade-off between

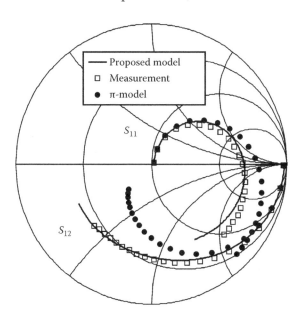

Figure 3.14 Comparison between measured and simulated S-parameters of the E3.5 inductor. (From A. Scuderi et al., 2004, "A Lumped Scalable Model for Silicon Integrated Spiral Inductors," *IEEE Transactions on Circuits and Systems Part I*, vol. 51, pp. 1203–1209. © 2004 IEEE. With permission.)

one- and two-port behavior is achieved by using $K = 0.16$, as will be fully verified in the following.

Accurate prediction of the Q requires a proper modeling of the inductor series resistance, R_S. Indeed, as frequency increases the effective resistance of the spiral rises due to the skin and proximity effects. The result of both phenomena produces the well-known current crowding, which can hardly be modeled in spiral geometries by using analytical equations. As a consequence, the empirical law reported in the following equation was formulated and its parameters extracted from experimental measurements of the real part of $-1/Y_{12}$, which provided the series resistance up to the peak quality factor frequency:

$$R_S = R_{DC} \cdot \frac{1 + \dfrac{f}{f_0} + \left(\dfrac{1.7f}{f_{SR}}\right)^2}{1 + \left(\dfrac{f}{f_{SR}}\right)^2} \tag{3.8}$$

where R_{DC} is the dc resistance of the spiral and f_{SR} is calculated using

$$f_{SR} = \frac{1}{\left(2\pi \cdot \sqrt{L_S \cdot C_{OX}}\right)} \tag{3.9}$$

In (3.8) the value of f_0 was set to 7 and 3.5 GHz for widths equal to 6 and 10 μm, respectively. It was set to 2.8 GHz for 14, 18, and 20 μm.

The classical one- and two-dimensional approximations of R_S (Rs_{1D} and Rs_{2D}, respectively) were employed in the proposed topology for comparison with (3.8).

$$R_{s\,1D} = R_{DC} \cdot \frac{t}{\delta \cdot \left(1 - e^{-t/\delta}\right)} \tag{3.10}$$

$$R_{s\,2D} = R_{s\,1D} \cdot \frac{1}{\left(1 + \dfrac{t}{w}\right)} \tag{3.11}$$

where t and w are the thickness and width of the metal (Eo and Eisenstadt 1993).

Equation (3.8) was used in the proposed and standard π-models to highlight the enhancement deriving from the novel topology. The simulated and measured equivalent series resistances of the E3.5 inductor, normalized with respect to the dc value, are shown in Figure 3.15. It is clear that equation (3.8) in the proposed model provides the best fit to measured data. This helps to demonstrate the validity of the split topology, because using the same equation in a standard π-model does not provide a correct estimation of the equivalent series resistance at higher frequencies.

Substrate effects were taken into account by two oxide capacitances (C_{OX}) in series with the RC networks that model the PGS. It is well known that the displacement of charge between the spiral and buried layer determines capacitive effects that have a large impact on both the f_{SR} and the Q. C_{OX} arises from both area (C_A) and perimeter (C_P) effects, which can be easily estimated with equations (3.12) and (3.13), respectively:

$$C_A = A \cdot \frac{\varepsilon_{ox}}{t_{ox}} \tag{3.12}$$

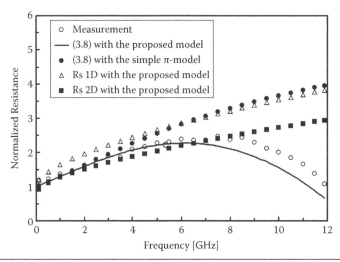

Figure 3.15 Simulated and measured normalized equivalent series resistance of the E3.5 inductor. (From A. Scuderi et al., 2004, "A Lumped Scalable Model for Silicon Integrated Spiral Inductors," *IEEE Transactions on Circuits and Systems Part I*, vol. 51, pp. 1203–1209. © 2004 IEEE. With permission.)

where A is the area of the spiral and ε_{ox} and t_{ox} are the oxide dielectric permittivity and thickness, whose value was extracted from the experimental measurements:

$$C_P = P \cdot C_{SP} \tag{3.13}$$

where P is the length of spiral perimeter and C_{SP} is the per-unit-length specific capacitance. The simulated constant E field contours reveal that the main perimeter capacitive contribution is due to both the inner and outer circumferences, whereas the electrical coupling between adjacent turns can be neglected to a first order. The total oxide capacitance was calculated using equation (3.14):

$$C_{OX} = \frac{1}{2} \cdot (C_A + C_P) \tag{3.14}$$

The capacitive effect due to overlap between the spiral and the underpass is modeled by the capacitor C_S whose value was calculated by equations (3.12) and (3.13), where t_{ox} and C_{SP} are the thickness and capacitance per-unit-length between spiral and underpass, respectively.

The radial patterned ground shield can be easily modeled by a RC network. The value of R_G was calculated from the geometrical parameters of each circular sector according to equation (3.15):

$$R_G = \frac{1}{N_{set}} \cdot \frac{R_{sheet} \cdot L_{PGS}}{W_{PGS}} \tag{3.15}$$

where R_{sheet} is the buried layer sheet resistance, N_{set} is the number of sectors, and L_{PGS} and W_{PGS} are the equivalent length and width of each sector, respectively. Capacitor C_G was calculated by using the well-known silicon time constant formula $R_G\, C_G = \rho_{Si}\, \varepsilon_{Si}$ (Pfost, Rein, and Holztwarth 1996), where ρ_{Si} and ε_{Si} are the resistivity and dielectric permittivity of the buried layer.

In Figure 3.13, L_S represents inductive behavior whose value can be calculated by referring to the consideration and formulas reported in Section 3.4.

3.5.2 Model Validation

In this section, the proposed model was validated by comparisons with fully de-embedded two-port measurements of integrated inductors for both single-ended and differential excitation, whose layout details are reported in Section 3.3. Because L_S is calculated by current sheet approximation, separate discussion will be done for sub-nH inductors in order to highlight the soundness of the model when using the corrected equation (3.3) instead of (3.1). The inductance and Q were used as performance parameters to validate the accuracy of the model with one terminal grounded, and S-parameters were employed for the two-port behavior.

Figure 3.16 and Figure 3.17 show simulations and measurements of inductance, Q- and S-parameters of several asymmetric inductors with different geometrical parameter.

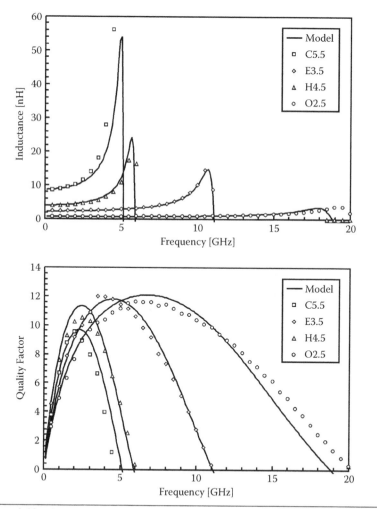

Figure 3.16 Comparisons between measured (symbols) and simulated quality factors (solid lines) for single-ended inductors.

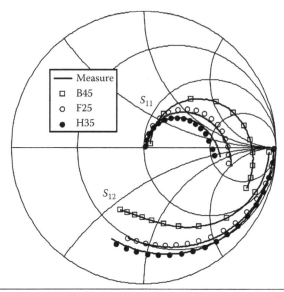

Figure 3.17 Comparisons between measured (symbols) and simulated S-parameters (solid lines) for single-ended inductors.

The correct estimation of the f_{SR} confirms the validity of the capacitance contribution calculations carried out using equations (3.12) and (3.13). Moreover, thanks to both the novel equation for the series resistance and the proposed topology, Q was accurately predicted over the whole frequency range. Close agreement was also found for the peak Q, which is of utmost importance when optimized inductors have to be designed for a given operating frequency. Moreover, Figure 3.17 confirms the accuracy of the proposed topology regarding the two-port behavior as mentioned in Section 3.5.1. Maximum errors on S_{11} and S_{12} are less than 1 dB and 5 degrees on magnitudes and phases, respectively. These results represent a substantial improvement compared to common lumped models, where simulations are usually compared in terms of only one-port equivalent inductance and Q.

In Figure 3.18 the proposed model is also compared with experimental data of symmetric inductors. It is worth noting that symmetrical coils are not used to extract equations of the model. Errors lower than 8 and 4% are achieved for Q and f_{SR}, respectively, in symmetric inductors covering inductance values from 0.2 to 7 nH. The soundness of the estimation, provided by the model, points out the validity of both topology and model equations for series and substrate losses.

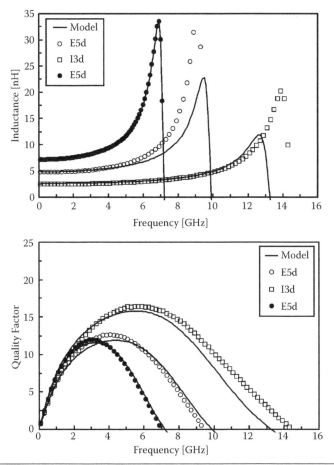

Figure 3.18 Comparisons between measured (symbols) and simulated quality factors (solid lines) for symmetric inductors.

In the case of sub-nH asymmetric inductors both original current sheet equations (3.1) and modified one (3.3) have been used to calculate the series inductance in the lumped scalable model. Because (3.1) and (3.3) only provide the self-inductance of the spiral alone, the underpass inductance was calculated using expressions given in Greenhouse (1974). The weak magnetic coupling with the substrate and ground plane was taken into account by reducing the series inductance according to

$$L_{\text{eff}} = \left(L_{\text{sp}} + L_{\text{up}}\right)\cdot\left(1 - k^2\right) \tag{3.16}$$

where L_{eff} is the effective inductance that takes into account the magnetic coupling, L_{sp} is the spiral inductance calculated using (3.1) or (3.3), L_{up}

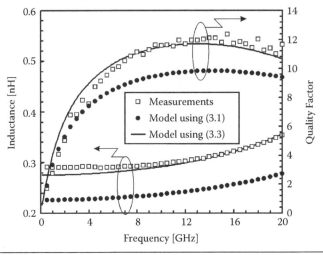

Figure 3.19 Comparison between measurements and model simulations using (3.1) and (3.3) for a 0.3-nH inductor ($n = 1.5$, $w = 20$ µm, $d_{in} = 50$ µm). (Modified from T. Biondi et al., *IEEE Microwave and Wireless Components Letters,* 15: 922–924, 2005, © 2005 IEEE. With permission.)

is the underpass inductance calculated using Greenhouse (1974) and k (=0.12) is the magnetic coupling coefficient estimated by means of proper EM simulations. Lumped model simulations and measurements of 0.3 and 0.5 nH inductors are compared in Figure 3.19. Results show that (3.1) underestimates inductance and Q with errors higher than 20%, in agreement with data reported in Figure 3.19. On the other hand, the correct value of the inductance calculated by (3.3) allows an excellent estimation of the Q with substantial error reduction as compared to (3.1).

The scalability of the model over the whole range of geometrical parameters is further confirmed by Figure 3.20, where the error distribution curves of inductance, f_{SR}, and Q at four working frequencies are reported. Thanks to the accuracy of current sheet approximation the inductance can be estimated with median and maximum errors of 2 and 7.2%, respectively, for all integrated inductors. Moreover, the error distribution curves of Q highlight the soundness of both the topology and closed forms describing series and substrate losses. Indeed, in more than 50% of measured devices the Q is predicted with error lower than 7% for working frequencies up 20 GHz. The accuracy of the predicted Q at 2 and 5 GHz, which features maximum errors lower than 14%, indicates that this model is a powerful optimization tool for integrated inductors for applications operating in these bands.

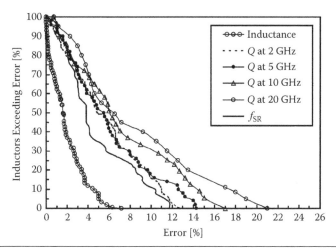

Figure 3.20 Error distributions of low-frequency inductance, f_{SR}, and quality factor at four working frequencies. (From A. Scuderi et al., 2006, "A 18-GHz Silicon Bipolar VCO with Transformer-Based Resonator," *IEEE RFIC Symposium Technical Digest*, pp. 491–494. © 2006 IEEE. With permission.)

Moreover, the f_{SR} was estimated to have maximum and average errors of 12% and less than 5%, respectively.

References

G. Avenier, M. Diop, P. Chevalier, et al., 2009, "0.13 μm SiGe BiCMOS Technology Fully Dedicated to mm-Wave Applications," *IEEE Journal of Solid-State Circuits*, vol. 44, pp. 2312–2321.

T. Biondi, A. Scuderi, E. Ragonese, and G. Palmisano, 2005, "Sub-nH Inductor Modeling for RF IC Design," *IEEE Microwave and Wireless Components Letters*, vol. 15, pp. 922–924.

Y. S. Choi, and J. B. Yoon, 2004, "Experimental Analysis of the Effect of Metal Thickness on the Quality Factor in Integrated Spiral Inductors for RF ICs," *IEEE Electron Device Letters*, vol. 25, pp. 76–79.

D. Coolbaugh, E. Eshun, R. Groves, et al., 2002, "Advanced Passive Devices for Enhanced Integrated RF Circuit Performance," *IEEE RFIC Symposium Technical Digest*, pp. 341–344.

M. Danesh, and J. R. Long, 2002, "Differentially Driven Symmetric Microstrip Inductors," *IEEE Transactions on Microwave Theory and Techniques*, vol. 50, pp. 332–341.

Y. Eo, and W. R. Eisenstadt, 1993, "High-Speed VLSI Interconnect Modeling Based on S Parameter Measurements," *IEEE Transactions on Components, Hybrids, and Manufacturing*, vol. 16, pp. 555–562.

H. M. Greenhouse, 1974, "Design of Planar Rectangular Microelectronic Inductors," *IEEE Transactions on Parts, Hybrids, and Packaging*, vol. 10, pp. 101–109.

S. Jenei, B. K. J. C. Nauwelaers, and S. Decoutere, 2002, "Physics Based Closed-Form Inductance Expression for Compact Modelling of Integrated Spiral Inductors," *IEEE Journal of Solid-State Circuits*, vol. 37, pp. 77–80.

H. Knapp, B. Dehlink, H-P Forstner, et al., 2007, "SiGe Circuits for Automotive Radar," *IEEE Silicon Monolithic Integrated Circuits in RF Systems*, pp. 231–236.

S. R. Kythakyapuzha, and W. B. Kuhn, 2001, "Modeling of Inductors and Transformers," *IEEE RFIC Symposium Technical Digest*, pp. 283–286.

H. Li, B. Jagannathan, J. Wang, et al., 2007, "Technology Scaling and Device Design for 350 GHz RF Performance in a 45nm Bulk CMOS Process," *IEEE Symposium on VLSI Technology Digest of Technical Papers*, pp. 56–57.

J. R. Long, and M. A. Copeland, 1997, "The Modeling, Characterization, and Design of Monolithic Inductors for Silicon RF IC's," *IEEE Journal of Solid-State Circuits*, vol. 32, pp. 357–368.

D. Melendy, P. Francis, C. Pichler, K. Hwang, G. Srinivasan, and A. Weisshaar, 2002, "A New Wide Band Compact Model for Spiral Inductors in RFICs," *IEEE Electron Devices Letters*, vol. 23, pp. 273–275.

S. S. Mohan, 1999, *The Design, Modeling and Optimization of On Chip Inductor and Transformer Circuits*, Ph.D. thesis, Stanford University, CA.

S. S. Mohan, M. del Mar Hershenson, S. P. Boyd, and T. H. Lee, 1999, "Simple Accurate Expression for Planar Spiral Inductances," *IEEE Journal of Solid-State Circuits*, vol. 34, pp. 1419–1424.

M. Pfost, H. M. Rein, and T. Holztwarth, 1996, "Modeling Substrate Effects in the Design of High Speed Si Bipolar IC's," *IEEE Journal of Solid-State Circuits*, vol. 31, pp. 1493–1501.

E. Ragonese, A. Scuderi, T. Biondi, and G. Palmisano, 2004, "Experimental Comparison of Substrate Structures for Inductors and Transformers," *IEEE MELECON Symposium Technical Digest*, pp. 143–146.

A. Scuderi, T. Biondi, E. Ragonese, and G. Palmisano, 2004, "A Lumped Scalable Model for Silicon Integrated Spiral Inductors," *IEEE Transactions on Circuits and Systems Part I*, vol. 51, pp. 1203–1209.

A. Scuderi, T. Biondi, E. Ragonese, and G. Palmisano, 2005, "Analysis and Modeling of Thick Metal Spiral Inductors on Silicon," *Proceedings of the European Microwave Conference*, pp. 81–84.

A. Scuderi, E. Ragonese, T. Biondi, and G. Palmisano, 2006, "A 18-GHz Silicon Bipolar VCO with Transformer-Based Resonator," *IEEE RFIC Symposium Technical Digest*, pp. 491–494.

C. P. Yue, and S. S. Wong, 1998, "On-Chip Spiral Inductors with Patterned Ground Shields for Si-Based RF ICs," *IEEE Journal of Solid-State Circuits*, vol. 33, pp. 743–752.

C. P. Yue, and S. S. Wong, 2000, "Physical Modeling of Spiral Inductors on Silicon," *IEEE Transactions on Electron Devices*, vol. 47, pp. 560–568.

4

ANALYSIS AND MODELING OF SILICON-INTEGRATED TRANSFORMERS

Nowadays, on-chip transformers are widely used to implement functions such as impedance conversion, resonant loads, low-noise feedback in amplifiers, bandwidth enhancement, and differential-to-single conversion (Zhou and Allstot 1998; Cassan and Long 1999; Long 2000; Bhatti, Roufoogaran, and Castaneda 2005; Kwok and Luong 2005). Because the amount of silicon area occupied by transformers can be a limiting factor in most applications, interleaved or tapped structures are often replaced by stacked configurations, which offer higher magnetic coupling and area efficiency, although at the expense of increased parasitic capacitances. Patterned ground shields can be profitably exploited in all transformer configurations to reduce losses caused by eddy currents flowing into the substrate (Niknejad and Meyer 2001; Ng, Rejaei, and Burghartz 2002).

Monolithic transformers are often fabricated by using coupled inductors. Although different layouts can be used to maximize the coupling between coils, only two types provide acceptable performance: stacked and interleaved transformers. Stacked transformers (Zolfaghari, Chan, and Razavi 2001) are made up using two identical coils fabricated using different metal layers placed one on top of the other, as shown in Figure 4.1. Interleaved transformers are built by two spirals fabricated using the same metal layer whose coils are laterally interleaved, as shown in Figure 4.2. The former topology allows higher coupling factors, but coils have different electrical parameters because they are built with different metal layers. Interleaved structures provide full symmetry between two coils but poor magnetic coupling.

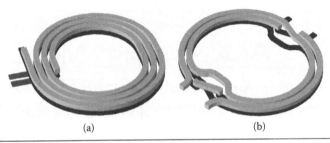

Figure 4.1 (a) Asymmetric and (b) symmetric stacked transformers.

Figure 4.2 Layout example of interleaved transformer.

The advantages of differential structures, discussed in Chapter 3 for monolithic inductors, can be easily extended to transformers. Examples of stacked and interleaved transformers for differential circuits are shown in Figure 4.1(b) and Figure 4.3, respectively.

In this chapter, the analysis and modeling of monolithic stacked transformers on silicon is addressed. First, the cross-section analysis carried out in Chapter 3 for monolithic inductors is extended to monolithic transformer in Section 4.1. The effect of layout scaling on self-resonance frequency, magnetic coupling coefficient (k), and insertion loss (S_{21}) is explored in Section 4.2 through on-wafer experimental measurements of several transformers with different geometries. Finally, a wideband lumped scalable model of integrated stacked transformers is reported in Section 4.3.

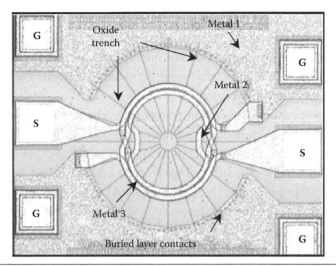

Figure 4.3   Fully symmetric transformer on buried layer PGS. (From E. Ragonese et al., 2004, "Experimental Comparison of Substrate Structures for Inductors and Transformers," *IEEE MELECON Symposium Technical Digest*, pp. 143–146. © 2004 IEEE. With permission.)

4.1 Substrate Optimization of Transformer on Silicon

The performance of transformers is strongly limited by the resistive losses in the metal of spirals and underlying substrate. Unconventional processes, such as SOI and MEM, are currently being investigated to overcome substrate effects. Despite their excellent performance, these solutions are expensive to comply with the low-cost requirements of the wireless market. Well-established silicon technologies provide a cost-competitive alternative for high-volume production. However, silicon-based VLSI processes typically employ highly conductive substrates, which considerably affect the transformer performance. Therefore a proper optimization procedure is required when designing the transformer.

Following the same considerations presented for monolithic inductors in Section 3.2, a comparative analysis on substrate vertical structure is carried out for stacked transformers (Ragonese et al. 2004). With respect to the analysis carried out for inductors, the Metal 1 PGS was not adopted because it suffers from very low self-resonance frequency (f_{SR}). To this aim, a circular transformer ($w = 8$ µm, $n = 2$, $s = 3$ µm, and $d_{in} = 210$ µm) was fabricated on a honeycomb oxide–trenched buried layer, solid buried layer, and buried layer PGS. The micrograph of the latter device is shown in Figure 4.3. Such transformer features a fully symmetric coil structure and, thanks to proper layout arrangements, the coils exploit only two metal layers, avoiding the more resistive bottom metal layer. Similar to inductors, also in this case, an oxide trench is used to hamper planar conduction in the highly doped n⁺ buried layer.

The comparison is based on transformer characteristic resistance (*TCR*), which was defined in Italia et al. (2005) as a suitable figure of merit related to the maximization of the available output power in tuned-load RF circuits. In Chapter 5 further details about this figure of merit will be provided. The transformer on buried layer PGS reveals the best *TCR*, whereas the device on solid buried layer provides the worst case because of induced eddy currents within the buried layer, as shown in Figure 4.4.

4.2   Transformer Characterization and Analysis

A wide set of stacked transformers on buried layer PGS was fabricated using the silicon technology detailed in Section 3.1.2. Metal 3 and Metal 2 were used for the primary and secondary coils, respectively, whereas the first metal layer was used for both the underpass and ground plane. A buried layer ground shield was built by exploiting a radial oxide trench pattern and connected to the ground plane

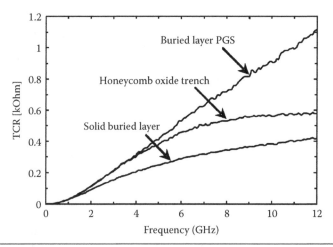

Figure 4.4   TCR comparison. (From A. Italia et al., 2005, "The Transformer Characteristic Resistance and Its Application to the Performance Analysis of Silicon Integrated Transformers," *IEEE RFK Symposium Technical Digest*, pp. 597–600. © 2005 IEEE. With permission.)

by a large number of buried layer contacts. Fabricated transformers had metal width from 6 to 20 µm, inner diameter from 50 to 150 µm, and turn number from 1.5 to 5.5. The intermetal spacing and ground plane distance were set to 4 and 50 µm, respectively. De-embedding structures were also fabricated and measured to improve the accuracy of high-frequency experimental data using the five-step de-embedding technique described in Chapter 2. A microphotograph of an integrated transformer is shown in Figure 4.5.

Figure 4.6 depicts the self-resonance frequency (f_{SR}) of the primary and secondary coils as a function of the low-frequency inductance (Biondi et al. 2006). Because the secondary coil was fabricated using the second metal layer (only 1.8 µm far from the substrate), the maximum f_{SR} does not exceed 35 GHz. This is considerably lower than that of inductors reported in Chapter 3 that were fabricated using the third metal layer of the same technology, which is 3.55 µm far from the substrate. Indeed, due to the high magnetic coupling coefficient ($0.8 < k < 0.96$), the substrate parasitic capacitance associated with the second metal layer is transferred almost entirely to the primary coil

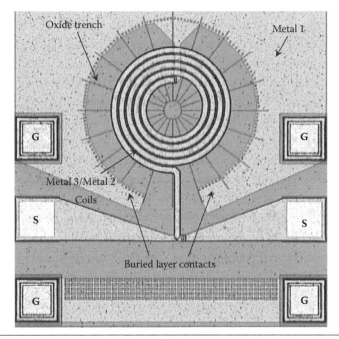

Figure 4.5 Micrograph of a buried layer PGS stacked transformer ($w = 10$ µm, $d_{in} = 100$ µm, $n = 4.5$).

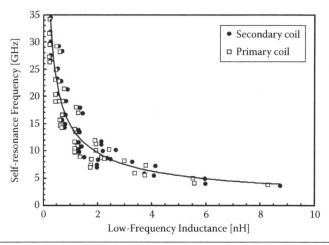

Figure 4.6 Self-resonance frequency versus low-frequency inductance for both coils. (From T. Biondi et al., 2006, "Analysis and Modeling of Layout Scaling in Silicon Integrated Stacked Transformers," *IEEE Transactions on Microwave Theory and Techniques*, vol. 54, pp. 2203–2210. © 2006 IEEE. With permission.)

that resonates at the same frequency as the secondary one, the turn ratio being approximately equal to unity.

Figure 4.7 shows the magnetic coupling coefficient as a function of frequency for different values of the metal width and for increasing turn number. Besides the increase with frequency, which is determined by the electric coupling between coils, enlarging the metal overlap by increasing the metal width and/or turn number causes magnetic coupling to rise. Indeed, as the area of the coils becomes larger, a greater percentage of the magnetic flux generated by one coil concatenates the other one, enhancing the coupling between them. This is further demonstrated by Figure 4.8, where the magnetic coupling coefficient improves for increasing outer diameters (d_{out}). Moreover, for a given outer diameter, higher values of k are obtained for smaller inner diameters. Based on these considerations, it can be concluded that magnetic coupling between the coils of stacked transformers can be enhanced by increasing the ratio d_{out}/d_{in}.

In spirals, larger outer diameters can be achieved by increasing the length of the conductor, which requires raising the turn number (indeed the effect of the metal width and intermetal spacing is much less important). Because in longer conductors both the series

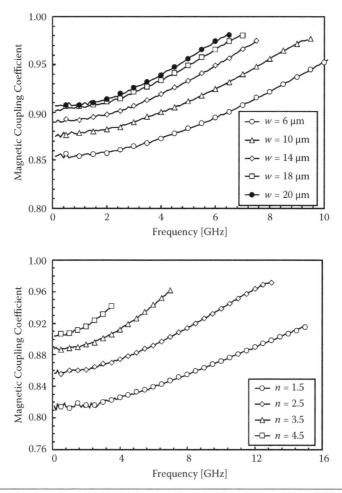

Figure 4.7   Magnetic coupling coefficient as a function of frequency for different metal widths ($d_{in} = 150$ μm, $n = 2.5$) and different turn numbers ($w = 10$ μm, $d_{in} = 100$ μm).

resistance and substrate capacitance are higher, more energy is dissipated along the metal trace or injected into the substrate. As a consequence, the overall gain of the transformer is reduced, although the magnetic coupling coefficient is higher. This can be observed from Figure 4.9, where the maximum value of the magnitude of S_{21} (in dB) decreases monotonically for increasing outer diameters. This effect is more pronounced in transformers with higher inner diameters in which adding one turn provides a higher relative increase of conductor length.

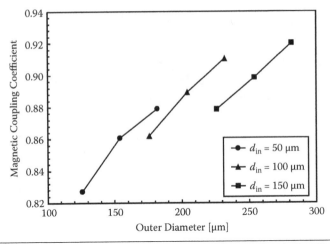

Figure 4.8 Magnetic coupling coefficient as a function of the outer diameter for different inner diameters ($w = 10$ μm).

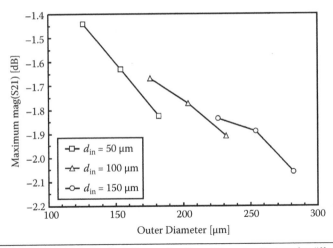

Figure 4.9 Maximum magnitude of S_{21} as a function of the outer diameter for different inner diameters ($w = 10$ μm).

4.3 Modeling of Stacked Transformers

The involved phenomena that take place in monolithic inductors make the lumped scalable modeling an issue of great challenge. The subject becomes even more complicated in monolithic transformers. In particular, in stacked structures the strong electromagnetic coupling between the two coils produces an involved series loss modeling. As a consequence, although many advances have been made in the fabrication of transformers on silicon, few significant results have been reported in the area of lumped circuit modeling. An analytical model using closed-form expressions was reported in Mohan et al. (1998). Simulation accuracy was verified by comparison with experimental measurements of a 20-nH patterned ground shield stacked transformer whose coils were shifted (both laterally and diagonally) with respect to each other. A uniform compact model for inductors and transformers was proposed in Long and Danesh (2001). Equivalent circuit components were calculated using numerical methods. Simulations were found to be in close agreement with measured data over a wide frequency range.

In this section, a wideband lumped scalable model for integrated stacked transformers is reported. Model components are calculated with closed-form expressions that make use of geometrical and technological data. Excellent agreement was found between simulated and measured S-parameters, coil inductance, magnetic coupling coefficient, and maximum available gain over a wide range of layout geometries.

4.3.1 Lumped Scalable Model

A model for stacked transformer on PGS is sketched in Figure 4.10. It represents a trade-off between classical lumped topologies—that is, those obtained by applying a π-network to both the primary and secondary coils—and distributed models. Indeed, traditional topologies are easy to manage but can hardly be employed to model monolithic transformers at relatively high frequencies and over a wide range of geometries. On the other hand, distributed networks simulate the real physical nature of the device at the expense of increased complexity.

The series impedances of the primary and secondary coils are split into three branches (with $\alpha + \beta + \gamma = 1$) that account for the inductance and resistance contributions of the inner, middle, and outer turns of the spirals. Indeed, due to both the different length of each turn and current crowding effects, the series impedance of the spiral is not uniformly distributed along the length of the conductor. Because current crowding is greater in transformers with many inner turns, the values of the splitting factors α, β, and γ depend on the geometrical parameters, especially on the fill factor. However, the experimental analysis of more than 50 structures with different layout revealed that this

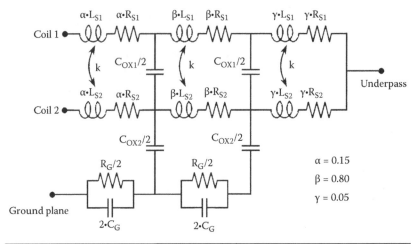

Figure 4.10 Wideband lumped model for PGS stacked transformers. (From T. Biondi et al., 2006, "Analysis and Modeling of Layout Scaling in Silicon Integrated Stacked Transformers," *IEEE Transactions on Microwave Theory and Techniques*, vol. 54, pp. 2203–2210. © 2006 IEEE. With permission.)

dependence is actually quite weak. As a consequence, layout independent coefficients were preferred for the sake of simplicity.

In the proposed model, the total inductance of the primary (L_{S1}) and secondary (L_{S2}) coils was calculated using the current sheet expression for circular spirals according to (4.1).

$$L_{S1,2} = \frac{\mu \cdot n^2 d_{avg}}{2} \cdot \left[\ln\left(\frac{2.46}{\rho}\right) + 0.2\rho^2 \right] \tag{4.1}$$

where $\rho = (d_{out} - d_{in})/(d_{out} + d_{in})$ is the fill factor.

Due to both skin and proximity effects, at a given frequency the current density inside the section of a conductor is not uniformly distributed but tends to crowd toward its outer surface. The cumulative effect of both these phenomena produces a frequency-dependent series resistance that is difficult to predict using physics-based expressions, especially in transformers where this phenomenon is further enhanced by the close proximity between the windings (Kuhn and Ibrahim 2001; Mayevskiy et al. 2005; Rotella et al. 2005). In the proposed model, the total series resistance of the spirals (R_{S1} and R_{S2}) takes both effects into account according to (4.2).

$$R_{S1,2} = R_{DC1,2} \cdot \left(1 + 0.1 \cdot \sqrt{f} + 0.002 \cdot f^2\right) \tag{4.2}$$

where $R_{S1,2}$ and $R_{DC1,2}$ are the ac and dc series resistances of the primary (subscript 1) and secondary (subscript 2) coils, respectively, and f is the frequency (expressed in GHz).

In stacked transformers, the amount of magnetic flux that couples one coil to the other depends on the geometrical parameters of the spirals (see Figure 4.7 to Figure 4.9). The magnetic coupling coefficient of investigated transformers varies from 0.7 to 0.92 as a result of the different layout. Due to this wide range of values, the geometry dependence of this parameter was explicitly taken into account in the proposed model. A closed-form expression that relates the transformer mutual inductance to layout parameters has only been developed for square geometries (Hsu 2005). On the other hand, the use of a constant magnetic coupling coefficient, commonly adopted in the literature, leads to systematic errors as large as 13%.

In the equivalent circuit of Figure 4.10, each branch of the primary coil is magnetically coupled to its respective branch of the secondary coil through the coefficient k. The dependence of k on the layout of the coils was taken into account using a monomial expression, as reported in (4.3).

$$k = x_0 \cdot n^{x_1} \cdot w^{x_2} \cdot d_{out}^{x_3} \tag{4.3}$$

where x_0, x_1, x_2, and x_3 are coefficients determined by least square fitting (4.3) to measured data. Their values were set to 0.46, 0.06, 0.03, and 0.10, respectively. The accuracy of (4.3) is reported in Table 4.1, which compares the measured and calculated magnetic coupling coefficients of transformers with different geometrical layout parameters.

Table 4.1 Comparison between Measured k and the Monomial Expression

w (μm)	n	d_{in} (μm)	k_{meas}	k_{model}	ERROR (%)
6	2.5	50	0.799	0.8	0.2
6	3.5	100	0.869	0.86	−1.08
6	5.5	150	0.919	0.92	0.01
10	2.5	50	0.827	0.828	0.12
10	3.5	100	0.889	0.885	−0.45
10	4.5	150	0.92	0.928	0.82
14	3.5	50	0.877	0.885	0.93
14	2.5	100	0.882	0.874	−0.84
14	2.5	150	0.891	0.894	0.33

Maximum and average errors of (4.3) with respect to measurements are around 1 and 0.5%, respectively.

Substrate effects were taken into account by means of oxide capacitances and RC networks, which model the radial patterned ground shield. The port-to-port (C_{OX1}) and port-to-substrate (C_{OX2}) capacitances, which arise from both area and perimeter contributions, were calculated using (4.4).

$$C_{OX1,2} = A \cdot \frac{\varepsilon_{ox1,2}}{t_{ox1,2}} + P \cdot C_{SP1,2} \tag{4.4}$$

where $\varepsilon_{ox1,2}$ and $t_{ox1,2}$ are the oxide dielectric permittivity and thickness, respectively; $C_{SP1,2}$ is the per-unit-length specific capacitance; A is the area of the spirals; and P is the length of the inner and outer circumferences of the coils. The impedance of the patterned ground shield was modeled through R_G and C_G, whose values were determined using expressions reported in Section 3.5.

4.3.2 Model Validation

The geometrical scalability of the model and its accuracy up to the transformer self-resonance frequency are demonstrated by comparison with experimental measurements in Figures 4.11 to Figure 4.14. Displayed data cover a wide range of layout geometries and coil inductances. The magnitude and phase of the simulated and measured S-parameters of two transformers with different turn number, metal width, and inner diameter are compared in Figure 4.11 and Figure 4.12 as a function of frequency. Displayed results reveal that the proposed model is in excellent agreement with measured data over a wide frequency range. This demonstrates that the topology adopted, despite its reduced complexity, provides a very good approximation of the distributed structure of transformers. Moreover, because the reported results involve devices with very different geometries, they prove the scalability of the model and confirm that it can be employed to predict the performance of monolithic stacked transformers over a wide range of layout parameters. To compare the capabilities of the proposed model with others referenced in the literature, some of the traditional figures of merit usually employed in the design of RF

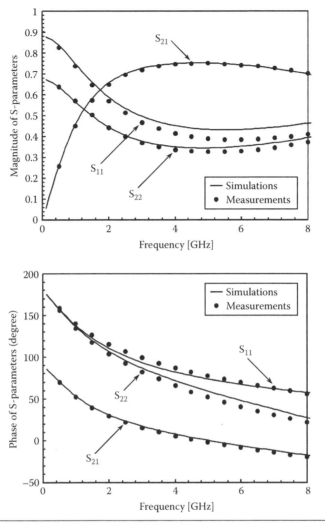

Figure 4.11   Magnitude and phase of simulated and measured S-parameters ($n = 5.5$, $w = 6$ µm, $d_{in} = 50$ µm). (From T. Biondi et al., 2006, "Analysis and Modeling of Layout Scaling in Silicon Integrated Stacked Transformers," *IEEE Transactions on Microwave Theory and Techniques*, vol. 54, pp. 2203–2210. © 2006 IEEE. With permission.)

ICs were also reported. Figure 4.13 reports the simulated and measured primary coil inductance, magnetic coupling coefficient, and maximum available gain (MAG) as a function of frequency up to the self-resonance frequency. The high degree of accuracy and geometrical scalability further demonstrate that the proposed solution can be profitably exploited in the design and optimization of RF circuit blocks. In the design of RF ICs, the f_{SR} is commonly employed as a

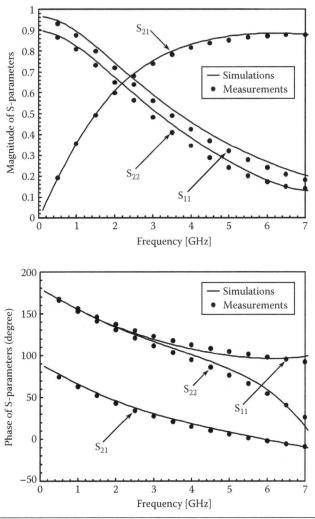

Figure 4.12 Magnitude and phase of the simulated and measured S-parameters ($n = 2.5$, w = 20 μm, d_{in} = 150 μm). (From T. Biondi et al., 2006, "Analysis and Modeling of Layout Scaling in Silicon Integrated Stacked Transformers," *IEEE Transactions on Microwave Theory and Techniques*, vol. 54, pp. 2203–2210. © 2006 IEEE. With permission.)

useful figure of merit to identify the bandwidth limitations introduced by inductive components. Indeed, it provides a direct estimation of the parasitic capacitances, which are of the utmost importance to predict the high-frequency dynamics of both inductors and transformers. This parameter is especially important in silicon technologies because the flow of currents into the semiconductive substrate causes much more frequency constrains than in GaAs or SOI-based

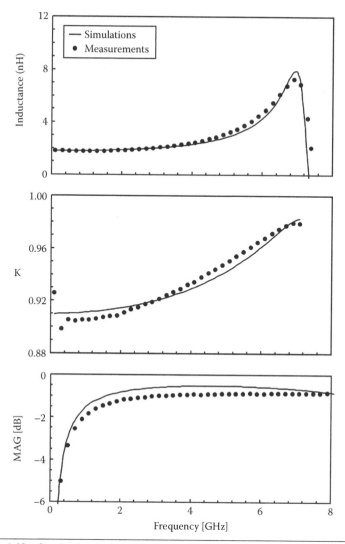

Figure 4.13 Simulated and measured primary inductance, k, and *MAG* as a function of frequency ($n = 2.5$, $w = 18$ μm, $d_{in} = 150$ μm). (From T. Biondi et al., 2006, "Analysis and Modeling of Layout Scaling in Silicon Integrated Stacked Transformers," *IEEE Transactions on Microwave Theory and Techniques*, vol. 54, pp. 2203–2210. © 2006 IEEE. With permission.)

ICs (Bahl 2001; Kelly and Wright 2002). Hence accurate estimation of the f_{SR} is a basic requirement of lumped scalable models. Figure 4.14 reports the error distribution curve of self-resonance frequency, which presents median and maximum errors of 3 and 12.5%, respectively, for all integrated transformers. The very close agreement between simulations and measurements further demonstrates the

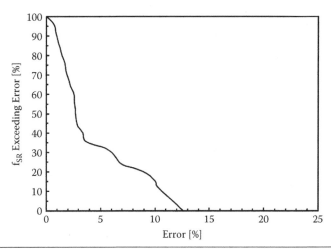

Figure 4.14   Exceeding error of self-resonance frequency.

suitability of the expressions used for capacitance calculations and the soundness of the proposed topology.

References

I. J. Bahl, 2001, "High-Performance Inductors," *IEEE Transactions on Microwave Theory and Techniques*, vol. 49, pp. 654–664.

I. Bhatti, R. Roufoogaran, and J. Castaneda, 2005, "A Fully Integrated Transformer-Based Front-End Architecture for Wireless Transceivers, *Digest of Technical Papers - IEEE International Solid-State Circuits Conference*, pp. 106–107.

T. Biondi, A. Scuderi, E. Ragonese, and G. Palmisano, 2006, "Analysis and Modeling of Layout Scaling in Silicon Integrated Stacked Transformers," *IEEE Transactions on Microwave Theory and Techniques*, vol. 54, pp. 2203–2210.

D. J. Cassan, and J. R. Long, 1999, "A 1-V Transformer-Feedback Low-Noise Amplifier for 5-GHz Wireless LAN in 0.18-μm CMOS," *IEEE Journal of Solid-State Circuits*, vol. 38, pp. 427–435.

H.-M. Hsu, 2005, "Implementation of High-Coupling and Broadband Transformer in RFCMOS Technology," *IEEE Transactions on Electron Devices*, vol. 52, pp. 1410–1414.

A. Italia, F. Carrara, E. Ragonese, T. Biondi, A. Scuderi, and G. Palmisano, 2005, "The Transformer Characteristic Resistance and Its Application to the Performance Analysis of Silicon Integrated Transformers," *IEEE RFIC Symposium Technical Digest*, pp. 597–600.

D. Kelly, and F. Wright, 2002, "Improvements to Performance of Spiral Inductors on Insulator," *Digest - IEEE MTT-S International Microwave Symposium*, pp. 541–543.

W. B. Kuhn, and N. M. Ibrahim, 2001, "Analysis of Current Crowding Effects in Multiturn Spiral Inductors," *IEEE Trans. Microwave Theory and Techniques*, vol. 49, pp. 31–38.

K. Kwok, and H. C. Luong, 2005, "Ultra-Low-Voltage High-Performance CMOS VCOs Using Transformer Feedback," *IEEE Journal of Solid-State Circuits*, vol. 40, pp. 652–660.

J. R. Long, 2000, "Monolithic Transformers for Silicon RF IC Design," *IEEE Journal of Solid-State Circuits*, vol. 35, pp. 1368–1382.

J. R. Long, and M. Danesh, 2001, "A Uniform Compact Model for Planar RF/ MMIC Interconnects, Inductors and Transformers," *Proceedings of the IEEE Bipolar/BiCMOS Circuits Technology Meeting*, pp. 167–170.

Y. Mayevskiy, A. Watson, P. Francis, K. Hwang, and A. Weisshaar, 2005, "A New Compact Model for Monolithic Transformers in Silicon-Based RFICs," *IEEE Microwave and Wireless Components Letters*, vol. 15, pp. 419–421.

S. S. Mohan, C. P. Yue, M. del Mar Hershenson, S. S. Wong, and T. H. Lee, 1998, "Modeling and Characterization of On-Chip Transformers," *IEEE International Electron Devices Meeting*, pp. 531–534.

K. T. Ng, B. Rejaei, and J. N. Burghartz, 2002, "Substrate Effects in Monolithic RF Transformers on Silicon," *IEEE Transactions on Microwave Theory and Techniques*, vol. 50, pp. 377–383.

A. M. Niknejad, and R. G. Meyer, 2001, "Analysis of Eddy-Current Losses over Conductive Substrates with Applications to Monolithic Inductors and Transformers," *IEEE Transactions on Microwave Theory and Techniques*, vol. 49, pp. 166–176.

E. Ragonese, A. Scuderi, T. Biondi, and G. Palmisano, 2004, "Experimental Comparison of Substrate Structures for Inductors and Transformers," *IEEE MELECON Symposium Technical Digest*, pp. 143–146.

F. Rotella, C. Cismaru, Y. Tkachenko, et al., 2005, "A Broad-Band Lumped Element Analytic Model Incorporating Skin Effect and Substrate Loss for Inductors and Inductor Like Components for Silicon Technology Performance Assessment and RFIC Design," *IEEE Transactions on Electron Devices*, vol. 52, pp. 1429–1441.

J. J. Zhou, and D. J. Allstot, 1998, "Monolithic Transformers and Their Applications in a Differential CMOS RF Low-Noise Amplifier," *IEEE Journal of Solid-State Circuits*, vol. 32, pp. 2020–2027.

A. Zolfaghari, A. Chan, and B. Razavi, 2001, "Stacked Inductors and Transformers in CMOS Technology," *IEEE Journal of Solid-State Circuits*, vol. 36, pp. 620–628.

5

Design Guidelines and Circuit Design Examples for Inductors and Transformers

This chapter provides suggestions and guidelines for optimum design of monolithic inductors and transformers. Hereinafter, the reference processes are silicon-based technologies—that is, CMOS and BiCMOS —while inductive components fabricated on dielectric substrates will be discussed in the Chapter 6. The effects of design parameters on the most important figures of merit of inductive components are commented, highlighting trends that can be exploited for performance enhancement. Because different approaches are adopted according to the application the device is being designed for, specific circuit examples along with the reference figures of merit and design procedures are reported. The chapter deals with several applications, such as 5-GHz WLAN, 17-GHz ISM communication, 24-GHz automotive radar sensor, and different functional RF/mm-wave blocks—that is, down-converter, voltage-controlled oscillator, transmitter—in order to highlight main optimization strategies for both inductors and transformers.

5.1 Application of Inductive Components in RF/mm-Wave ICs

In modern wireless applications integrated inductive components are essential and irreplaceable elements of a transceiver, especially at high frequencies. In particular, filtering capabilities, as well as noise, linearity, and power consumption performance, strongly depend on the availability of high-quality inductors and low-loss transformers. Moreover, they allow attaining a higher level of integration and lower costs by reducing and even removing expensive and bulky external components on the application boards.

Handling inductors and transformers may appear rather simple in comparison with transistors; however, maximizing their performance in CMOS and BiCMOS technologies is a difficult task for RF/mm-wave circuit designers, due to a widespread lack of reliable automatic procedures for the device optimization within a circuit block. As operative frequency increases the scenario becomes more involved. Indeed, on the one hand the availability of well-characterized inductive components considerably reduces in common design kits; on the other hand, the influence of passive components on the overall circuit performance greatly increases. As previously said in Chapter 2, a good knowledge of main loss mechanisms is a *condicio sine qua non* to manage the optimization procedure of both inductors and transformers. However, because there are several applications of inductive components in RF/mm-wave ICs, it is also required to recognize the most suitable figures of merit to be optimized. The Q-factor maximization of a spiral at a given value of inductance is the most frequent problem, but it is far from exhausting design issues, especially for integrated transformers.

For instance, Figure 5.1 reports several circuit configurations, showing different applications of both inductors and transformers in RF/mm-wave ICs. In Figure 5.1(a) integrated inductor L_M is used along with capacitors C_{M1} and C_{M2} to implement a narrow-band impedance matching between transistors Q_1 and Q_2 in a multistage amplifier. Such reactive matching networks are largely employed in the most important transceiver building blocks in order to improve power, gain and noise performance. In Figure 5.1(b) we can see a classic configuration of an LNA. Simultaneous impedance and noise matching to the 50-ohm input source is obtained by means of emitter and base inductors, L_E and L_B, which are properly sized to tune the real and imaginary part of the LNA input impedance, respectively. In this circuit, power gain performance is optimized exploiting a tuned resonant load (C_{Load}, L_{Load}), which at the operating (resonance) frequency provides an equivalent resistance proportional to the ωQL product, without the drawbacks of a resistive load in terms of frequency bandwidth, voltage headroom, or noise (Girlando, Ragonese, and Palmisano 2004).

In Figure 5.1(c) a cross-coupled VCO is depicted, which makes use of a transformer-coupled LC tank. This topology allows maximizing both tuning range and phase noise performance, overcoming the

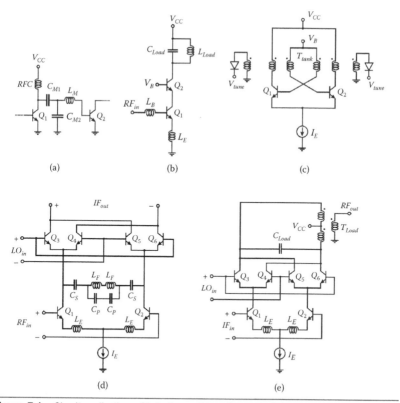

Figure 5.1   Circuit applications of inductors and transformers in RF/mm-wave ICs.

limitations of traditional LC configurations, as it will be explained in Section 5.4.2. In Figure 5.1(d) the filtering capabilities of integrated passive components are demonstrated in an image-reject down-converter. The double-balanced mixer includes a third-order differential notch filter (C_S, C_P, L_F) that provides on-chip image rejection. Integrated inductors are also used in the place of resistors as emitter degeneration to improve the linear input range of the voltage-to-current converter Q_1-Q_2. Finally, in Figure 5.1(e) the integrated transformer T_{LOAD} is employed as tuned load in an up-conversion mixer to maximize the delivered output power. This is a typical example of a transformer-loaded circuit, whose optimization procedure is based on the transformer characteristic resistance (TCR), which is a proper extension to the transformers of the ωQL product. Moreover, T_{LOAD} implements the differential-to-single-ended conversion of output RF signal.

These circuit examples give only a brief and not exhaustive overview on the myriad of applications for inductors and transformers. The following sections will provide a better insight by means of actual design examples of RF/mm-wave circuits to support reader understanding.

5.2   Inductor Design Guidelines and Optimization Procedures

The design of integrated inductors was widely discussed by several authors using extensive measurements, analytical studies, EM simulations, or equivalent circuit lumped modeling (Long and Copeland 1997; Koutsoyannopoulos and Papananos 2000; Burghartz and Rejaei 2003; Rotella et al. 2005). However, suggested design procedures reveal slightly involved for unskilled designers. This section deals with the performance optimization of spiral inductors by taking advantage of both experimental measurements and simulated results from the inductor model already presented in Chapter 3 with the aim of providing the reader with a few simple design concepts. The discussion is manly focused on the coil design, neglecting the issue of the substrate optimization. Indeed, as reported in the previous chapters, the best and widely adopted substrate arrangement for silicon-integrated inductive components consists of a conductive PGS, which can be built using metal, highly doped silicon, or polysilicon layers (C. P. Yue and Wong 1998).

To design an integrated inductor means to define its layout parameters; that is, the coil shape, the number of turn (n), the metal width (w), the metal spacing (s), the inner diameter (d_{in}), and the coil thickness (t). Some of these parameters can be easily chosen by the designer on the basis of simple considerations, partially reported in Chapter 2. In particular, circular or polygonal shapes are preferred to the square one due to their lower losses—that is, higher Q-factor—at a given inductance value. Moreover, the metal spacing is generally fixed to the closest value allowed by the process because tight spirals achieve better inductive coupling, which means higher inductance and Q-factor, whereas the increased fringing capacitance is negligible in comparison with the capacitance toward the substrate. Therefore, coil sizing means to set the remaining parameters (i.e., n, w, d_{in}, and, if possible, t) on the basis of the specific circuit constraints.

Common figures of merit for inductor design are the following: the inductance (L), the Q-factor (Q), the self-resonance frequency (f_{SR}), the ωQL product, and the area.

A first important consideration is that there is a relationship between L and maximum Q-factor (Q_{MAX}) on one side and the f_{SR} on the other side. Indeed, as L increases, integrated spirals exhibit lower Q_{MAX} and f_{SR} values. These trends can be clearly observed in Figure 5.2, which reports measured Q_{MAX} and f_{SR} as functions of inductance for several spirals fabricated in a silicon bipolar process. Similar curves can be observed in other silicon-based technologies with different back-end

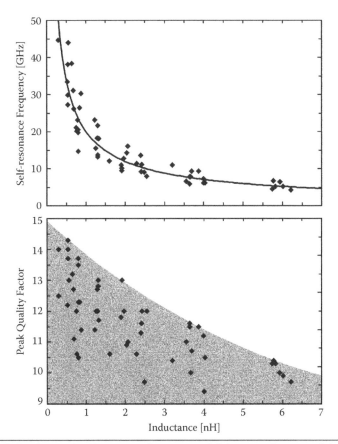

Figure 5.2 Measured self-resonance frequency and Q-factor performance of integrated inductors as function of inductance. (From A. Italia, L. La Paglia, A. Scuderi, F. Carrara, E. Ragonese, and G. Palmisano. "A Silicon Bipolar Transmitter Front-End for 802.11a and HIPERLAN2 Wireless LANs," *IEEE Journal of Solid-State Circuits,* vol. 40, pp. 1451–1459. © 2005 IEEE. With permission.)

of line (BEOL). Two simple comments can better clarify the meaning of Figure 5.2:

- The same L can be obtained with different f_{SR} (e.g., 1.3-nH spirals exhibit f_{SR} within the range 13–18 GHz, whereas for 2-nH spirals the differences in the f_{SR} can be wider, from 9.5 to 16 GHz).
- The same L can be obtained with different values of Q_{MAX} (e.g., 2.5-nH coils shows Q_{MAX} between 9.7 and 12).

A correct design of a spiral inductor starts from the definition of the f_{SR}, which limits the operating frequency range of the device. At first, a simple rule of thumb can be adopted by noting the fact that Q_{MAX} generally occurs at around one half the f_{SR}. Although Figure 5.2 indicates that for a fixed f_{SR} there is a limitation on the exploitable value of L (in a given technology), for the designer this constraint is actually less severe than it appears. Indeed, high-L spirals are adopted at low frequency, whereas high-frequency circuits usually require sub-nH inductors whose f_{SR} is not so limiting. Moreover, a proper choice of the spiral geometrical parameters generally allows the maximization of the f_{SR} at a given L.

A second consideration is that L is strictly related to the spiral length. As shown in Figure 5.3, at a fixed d_{in} the low-frequency inductance (L_{DC}) highly depends on the number of turns (according

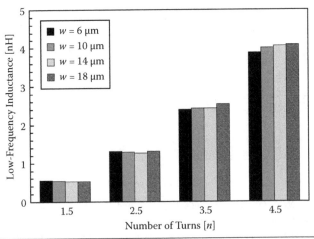

Figure 5.3 Low-frequency inductance of integrated spiral inductors as a function of the number of turns ($d_{in} = 100 \ \mu m$).

to a square law), whereas the inductance variation due to the metal width, w, is negligible. Therefore the key point is to set w in order to maximize Q and optimize the f_{SR} at a given L (which means at a given n). At low/medium frequencies the dominant losses of an integrated inductor are due to the series resistance of the spiral and large metals can be profitably used because they improve the linear slope of Q at the expense of a lower f_{SR}. On the other hand, at higher frequency both current crowding effects and the parasitic substrate capacitance force to reduce w. It is worth noting that the choice of w can also be influenced by the electromigration rules in "high-current" inductors; for example, in a power amplifier (PA), which can suffer from excessively large area and consequently low f_{SR}.

Figure 5.4 helps to better understand the effect of w on both Q and the f_{SR}. The diagram reports the Q frequency behavior of three 3.5-turn spirals using different metal widths at similar L_{DC} values (i.e., around 2.3 nH). As expected, the increase of the metal width produces a slope rise (at low frequency), a left shift of Q_{MAX}, and a reduction of the f_{SR}. A proper value of w can be found, which sets Q_{MAX} around the operative frequency (i.e., w = 8 μm at 5 GHz in the example reported in Figure 5.4), thus reducing the overall inductor losses.

For the sake of completeness, some comments concerning d_{in} are required. As explained in Chapter 2, the choice of d_{in} involves an accurate evaluation of the current crowding effects due to the magnetically

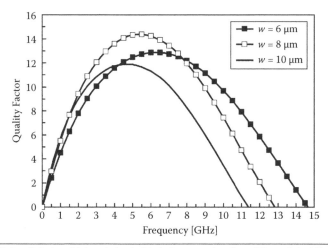

Figure 5.4 Q-factor of integrated spiral inductors as a function of the metal width ($n = 3.5$, $d_{in} = 100$ μm, $L_{DC} = 2.4$ nH).

induced currents in the inmost turns of the spiral. Indeed, it is well known that "hollow" inductors (i.e., with high d_{in} and low n) are preferred to "filled" inductors (i.e., with low d_{in} and high n) at the cost of larger device area (Craninckx and Steyaert 1997). On-wafer measurements of integrated inductors with different d_{in} and n, but with fixed w, s, and total length, which means at similar L_{DC}, demonstrate considerable degradation of Q for small core sizes (d_{in} = 10 μm) while indicating an optimal d_{in} value of 100 μm for typical spirals adopted in RF ICs (Sia et al. 2005). As a general rule of thumb, d_{in} lower than 50 μm should be avoided unless it is mandatory to obtain very low inductance values (L_{DC} < 0.5 nH), as it occurs for single-turn inductors usually adopted beyond 20 GHz. The fill ratio can be a precious figure of merit to discriminate hollow from filled spirals, using a value of 0.6 as a threshold (Mohan et al. 1999). However, the optimization of d_{in} with the aim of reducing the eddy currents can be actually carried out only by means of EM simulations, scalable models, or experimental measurements, because physics-based expressions or analytical methods are quite inaccurate.

The above considerations can be summarized in the following general design procedure:

- set a high enough d_{in} to limit the effects of eddy currents (by means of EM simulations);
- choose n to set the required range of L_{DC} (by using analytical inductance expressions);
- choose w to optimize Q by setting Q_{MAX} at the operative frequency (by using EM simulations or scalable lumped models);
- verify inductor performance and repeat the procedure if needed.

This optimization procedure well applies to the most frequent design constraint; that is, the maximization of Q at a fixed frequency and at a given L, as it is required in the design of inductive degeneration, matching network, etc. A different situation occurs when maximizing the ωQL product at the operative frequency is mandatory (e.g., when the inductor is used as resonant load). In this case, the design procedure involves both L and Q at the same time, because a proper combination of these two parameters must be found. The optimization strategies are based on maximizing either Q or L. Inductance maximization approach generally achieves better results because it is less difficult to increase L than Q. Indeed, low-L inductors can attain very high Q, but relative improvements on ωQL are negligible. On the other hand, high-L inductors can be designed with a good Q and relative improvements on ωQL are greater. The most profitable strategy consists of maximizing L at a given f_{SR} and then optimizing Q at the chosen value of L, according to the previously reported procedure. Obviously, the f_{SR} is a stringent specification, which determines the choice of both n and w, limiting the achievable values of the ωQL product. Fortunately, as the operative frequency increases, high ωQL products are also attainable at lower L. The design procedure has to take into account that the peak of the ωQL product (ωQL_{MAX}) generally occurs at a higher frequency than the Q_{MAX} and therefore the Q-factor maximization should be carried out at a lower frequency than the operating one. Figure 5.5 reports a typical relationship between Q_{MAX} and ωQL_{MAX} for an integrated spiral inductor (n = 3.5, w = 7 μm, s = 3 μm, d_{in} = 98 μm). In this example, it can be observed that Q_{MAX} and ωQL_{MAX} occur at 4.5 and 9.5 GHz, respectively, with f_{SR} as high as 11 GHz. This behavior (i.e., Q_{MAX} and ωQL_{MAX} occur at around $f_{SR}/2$ and f_{SR}, respectively) can be assumed as a general rule of thumb.

The reported design procedures have intentionally neglected one of the mentioned inductor parameters; that is, the metal thickness t. For a long time this parameter was hardly taken into account by designers,

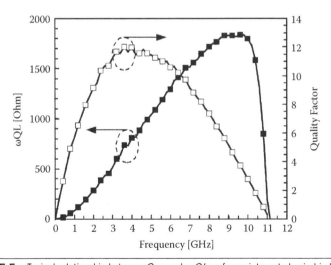

Figure 5.5 Typical relationship between Q_{MAX} and ωQL_{MAX} for an integrated spiral inductor.

because metals did not generally exceed 1-μm thickness in silicon technologies. Nowadays the benefits of thick metals on the performance of spiral inductors are becoming attractive, especially in modern CMOS and BiCMOS processes, which exploit improved BEOLs with at least two thick top metals (Avenier et al. 2009; Pastore et al. 2009). In this case, the use of multilayer structures, consisting of shunted metals (Burghartz, Soyuer, and Jenkins 1996), turns into an equivalent thickness higher than 7–8 μm. As already explained in Chapter 3, the increment of t produces a reduction of the L_{DC} value of a spiral inductor, which partially compensates for the benefit of the series resistance reduction. However, achievable improvements of the Q-factor are still appreciable, especially at low and medium frequencies. Some papers investigated the effects of metal thickness on the Q-factor of integrated spiral inductors by using experimental data, three-dimensional EM simulations, or physics-based modeling (Shiwei and Lihui 2003; Choi and Yoon 2004; Murphy et al. 2005; Scuderi et al. 2005). These studies clearly demonstrate that the benefits achievable by metal thickening are considerably reduced by the current crowding due to both skin and proximity effects, which are responsible for the saturation of the Q-enhancement at t higher than 10 μm and at increasing frequencies. In Choi and Yoon (2004) experimental measurements of spiral inductors with metal thickness from 5 to 22.5 μm indicate that the current crowding causes a saturation of the Q-enhancement at d_{in} of around 100 μm and/or at tight metal spacing (i.e., $s = 2$–3 μm). On the contrary, significant improvements can be further obtained using t beyond 10 μm if higher s and d_{in} are adopted. It is worth noting that the Q-enhancement is more pronounced at low and medium frequencies, far enough from the self-resonance of the inductors. These results are confirmed in the analysis reported in Scuderi et al. (2005), where a single-turn 16-μm-wide circular geometry with d_{in} as high as 240 μm was employed to investigate the described phenomena. Indeed, in this simple structure the proximity effect can be neglected, thus accounting for the skin effect only. Three-dimensional EM simulations of the described coil at three different metal thickness (i.e., 3, 9, and 15 μm) reveal that the Q-enhancement saturation occurs beyond 10 μm, because at $t = 15$ μm the performance variations with respect to the case of 3-μm-thick metal are around 62, 50, and 38% at 2, 5, and 12 GHz, respectively. This analysis points out that the proximity effect is the first cause of the saturation of the Q-factor at high metal thickness and suggests using increased spacing to avoid similar effects in multiturn inductors. Comparative simulations of 3.5-turn 10-μm-wide spirals with different metal spacing and thickness indicate that t/s ratio higher than 3 should be avoided.

From these considerations, some useful hints can be drawn to design a thick (or multilayer) spiral inductor without making high-t metal employment fruitless.

- If a multilayer structure is used, first evaluate the f_{SR} reduction due to the increased parasitic capacitance toward the substrate and its compatibility with the operating frequency.
- Adopt larger d_{in} (around 200 μm) and, for multiturn inductors, do not use very tight metal spacing (i.e., $t/s < 3$) to limit the performance degradation due to the proximity effects.

It is worth noting that all the above-reported design guidelines, although referred to single-ended spiral inductors, can be easily extended to symmetric inductors (Danesh and Long 2002), which are traditionally employed in differential circuit topologies thanks to their benefits in terms of f_{SR}, Q, and area, as discussed in Chapter 3.

Finally, as far as the use of integrated inductors in mm-wave circuits is concerned, further design guidelines are required. Thanks to the advances in silicon technologies (Chevalier et al. 2006), spiral inductors are profitably employed in the implementation of mm-wave fully integrated transceivers (Khanpour et al. 2008; Laskin et al. 2008; Parsa and Razavi 2009), providing great advantages in terms of area consumption compared with traditional transmission lines. The maximization of the f_{SR} and hence the minimization of parasitic capacitances is of utmost importance to work above 30 GHz (Dickson et al. 2005). The use of narrow metal ($w < 5$ μm) and small inner diameter ($d_{in} < 50$ μm) allows the reduction of the inductor footprint and therefore of the capacitance toward the substrate, whereas a wider spacing helps in decreasing the fringing capacitance, especially if thick metals (or multilayer structures) are exploited. Moreover, polysilicon or metal shields are usually avoided to further reduce the parasitic capacitance toward the substrate. Thanks to the low area of mm-wave inductors, the lack of a PGS has generally negligible effects on the substrate losses.

The last concern is related to the impact of dummy metal patterns that are automatically placed by computer-aided design tools to guarantee the chemical mechanical planarization (CMP) integrity in common sub-0.18-μm technologies. Although for inductors of around 150–200 pH the dummy placement around the spiral produces performance variations within the measurement accuracy, no dummy zones are usually defined manually by the designer to prevent unpredictable coupling effects, especially for very small components ($L < 100$ pH).

5.3   Transformer Design Guidelines and Optimization Procedures

Because integrated transformers are built by means of two magnetically coupled spirals, an unskilled reader might think that transformer design is a simple extension of inductor design. The above remark is not completely false because the optimization procedures discussed for spiral inductors maintain their validity as far as the drawing of stand-alone primary and secondary coils is concerned. However, this represents only the starting point for an optimum design of an integrated transformer because the impact of the EM couplings (i.e., electric and magnetic couplings) between windings has to be properly considered. In particular, the magnetic coupling between windings is measured by the magnetic coupling factor (k), which is typically around 0.7–0.9 for a monolithic transformer due to a poor confinement of the magnetic flux. As explained in Chapters 2 and 4, basic configurations to implement an integrated transformer are the interleaved and stacked ones (Long 2000). For the sake of clarity, main advantages and drawbacks of these two implementations are pointed out in Table 5.1. Interleaved configurations can be profitably used to implement symmetric windings with k as high as 0.7 by using the top metal available in the process, thus maximizing both the f_{SR} and the

Table 5.1   Advantages and Drawbacks of Interleaved and Stacked Configurations

CONFIGURATION	AREA	k	f_{SR}	CAPACITANCE	ELECTRICAL SYMMETRY	TURN RATIO
Interleaved	Medium	Medium	Excellent	Excellent	Excellent	Excellent
Stacked	Excellent	Excellent	Medium	Poor	Medium	Poor

Q-factor at the expense of the area consumption. On the other hand, stacked transformers achieve higher k and a better area exploitation but exhibit lower f_{SR} due to the increased port-to-port and substrate parasitic capacitances. Moreover, although primary and secondary spirals can be drawn geometrically identical, they are implemented on different metal layers, thus producing an electrical asymmetry in the transformer. This asymmetry is caused by the following two phenomena:

• Upper and lower spirals exhibit different series resistances and hence different Q-factors due to different values of the metal thickness.
• The parasitic capacitance toward the substrate differs for each winding because the upper spiral is electrically shielded from the substrate by the lower one.

Another important performance characteristic is related to the implementation of transformers with nonunitary turn ratios, which are profitably used as integrated baluns (i.e., balanced-to-unbalanced transformers) for impedance transformations or single-ended-to-differential conversion of RF signals. As described in Long (2000) and Rotella et al. (2006), 1:n and m:n integrated transformers can be easily implemented by using interleaved spirals in the top metal (with a slight degradation of k), whereas the stacked configuration is less suitable to build such transformers because it generally requires multiple metal layers and exhibits a significant reduction of both k and Q-factor.

Interesting modifications to these basic configurations were introduced to implement novel structures of monolithic transformers. Some of these implementations are worth noting in order to provide the reader with a complete state of the art of transformer design; for example, the single-turn, six-layer interlaced stacked transformer presented in Lin (2005) that claims a nearly perfect magnetic-coupling factor (i.e., $k = 1$) and the intercoil multilayer crossing structure proposed in Lim et al. (2008) that takes advantage of both lateral and vertical magnetic coupling, independently exploited in the traditional interleaved and stacked configurations, respectively.

In this complex scenario, the actual problem of an RF/mm-wave IC designer is identifying the most suitable figures of merit for an integrated transformer within a specific circuit application and hence using such figures of merit to compare different implementations in terms of substrate arrangement (i.e., different PGS structures, oxide patterns, etc.), layout configuration (i.e., interleaved, stacked, etc.), and geometrical parameters of primary and secondary coils (i.e., n, w, d_{in}, etc.). On the other hand, it is quite misleading to carry out a transformer evaluation on the basis of spiral performance parameters; that is, primary (L_1) and secondary inductance (L_2), primary (Q_1) and secondary Q-factor (Q_2), and k. Indeed, the most important measure for transformers in circuit applications is how well they are able to transfer power from the input to the output port. This power transfer is related not only to the transformer itself but to the impedances connected to its input and output, including the corresponding matching networks. In recent years, several authors proposed different approaches for the rating of integrated transformers, using as reference figures of merit either the insertion loss, IL (Long 2000), the maximum available gain, MAG (Ng, Rejaei, and Burghartz 2002), or the transformer characteristic resistance, TCR (Carrara et al. 2006). For the sake of completeness the expressions for IL, MAG, and TCR are reported herein:

$$IL = 20 \cdot \log |S_{21}| \tag{5.1}$$

$$MAG = \frac{S_{21}}{S_{12}} \cdot \left(K - \sqrt{K^2 - 1} \right) \tag{5.2}$$

$$TCR = \frac{4 \left| Z_{21} \right|^2 \operatorname{Re}\left\{ Z_{11}^* \Delta_Z \right\}}{\operatorname{Re}\left\{ \Delta_Z + Z_{11} Z_{22}^* \right\}^2 - \left| Z_{21} \right|^4} \tag{5.3}$$

where K and Δ_Z are given by following equations (5.4) and (5.5), respectively:

$$K = \frac{1 - \left| S_{11} \right|^2 - \left| S_{22} \right|^2 + \left| S_{11} \cdot S_{22} - S_{21} \cdot S_{12} \right|^2}{2 \cdot \left| S_{21} \right| \cdot \left| S_{12} \right|} \tag{5.4}$$

$$\Delta_Z = Z_{11} Z_{22} - Z_{12} Z_{21} \tag{5.5}$$

These reference figures of merit provide a useful performance characterization only if the transformer is operated under specified conditions at both the input (source) and output (load) ports. In particular, the *IL* and the *MAG* require 50-ohm input/output terminations and conjugate matching, respectively, whereas the *TCR* is the generalized extension of the ωQL product to transformers when operated as tuned load (i.e., the primary coil is driven by high-impedance transistor collectors/drains and conjugate impedance matching is provided between the secondary coil and the load). Both *IL* and *MAG* are proposed as performance parameters for the minimization of the transformer power loss. On the other hand, the *TCR* is related to the maximization of the circuit available output power and gain. These approaches are not equivalent. Indeed, the improvement of the *IL* (or *MAG*) does not always lead to an increase in the available output power and hence in the *TCR* and vice versa. A simple demonstration of how much the transformer evaluation depends on the choice of the figure of merit is given in Figure 5.6, which compares the *IL*, *MAG*, and *TCR* measured performance of two stacked transformers belonging to the set already used in Chapter 4. Adopted transformers have different geometrical parameters (Transformer A: $n = 3.5$, $w = 6$ µm, $d_{in} = 100$ µm, $s = 4$ µm; Transformer B: $n = 2.5$, $w = 20$ µm, $d_{in} = 50$ µm, $s = 4$ µm) and different electrical performance (Transformer A: $L_{DC1} = 2.44$ nH, $L_{DC2} = 2.7$ nH, $Q_{MAX1} = 9.9$, $Q_{MAX2} = 4.4$, $k = 0.87$, $f_{SR} = 10.1$ GHz; Transformer B: $L_{DC1} = 0.64$ nH, $L_{DC2} = 0.74$ nH, $Q_{MAX1} = 8.9$, $Q_{MAX2} = 5$, $k = 0.87$, $f_{SR} = 14.6$ GHz). It is quite evident that Transformer A achieves considerably higher *TCR* up to the self-resonance. However, the situation is reversed if the comparison is carried on in terms of *MAG*, because Transformer B exhibits better performance in the whole frequency range. Finally, adopting the *IL* as reference parameter would provide a third different reading, because the results of the comparison are frequency dependent. Similar transformer comparisons could be reported in terms of the substrate management techniques (Italia et al. 2005) or the layout configurations, but unambiguous conclusions can hardly be drawn.

From the above discussions it is quite evident that a proper choice of the transformer performance parameter is mandatory. Some simple hints can be suggested to the reader to match the figure of merit to the specific circuit application. The *TCR* is clearly the reference

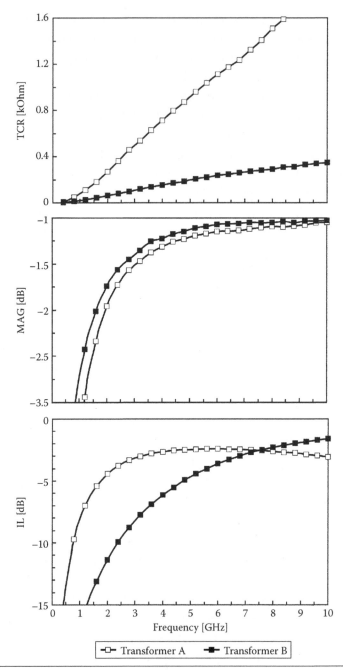

Figure 5.6 Comparison between two stacked transformers in terms of *IL*, *MAG*, and *TCR* (Transformer A: $n = 3.5$, $w = 6$ μm, $d_{in} = 100$ μm, $s = 4$ μm; Transformer B: $n = 2.5$, $w = 20$ μm, $d_{in} = 50$ μm, $s = 4$ μm).

performance parameter in transformer-loaded circuits, as the up-conversion mixer depicted in Figure 5.1(e). In this case, the transformer optimization procedure consists in maximizing the *TCR* (through proper choice of the both layout and substrate management), tuning out the transformer input inductance by means of a shunt capacitor, and, finally, providing conjugate impedance matching between the secondary coil and the load, as detailed in Carrara et al. (2006). The *IL* can be profitably adopted to optimize an integrated transformer when 50-ohm source and load terminations are employed, as it occurs when the transformer is included between the LNA and the antenna (e.g., to provide single-ended-to-differential conversion). In this case, it is well known that the *IL* also represents the noise figure of the transformer (Razavi 1998) and hence its minimization is of utmost importance for the overall performance of the *RX* chain. A common practice to reduce the *IL* is the transformer tuning by means of capacitors placed in shunt with the primary and secondary windings (Long 2000).

The use of the *MAG* is much more difficult to allocate because conjugate matching is infrequently provided at both primary and secondary coils. This parameter can be considered a general measure of the maximum power transfer efficiency of an integrated transformer.

Hereinafter, the discussion will be focused of the *TCR*, because transformer-loaded topologies are widely adopted in RF and mm-wave circuits (Simbürger et al. 1999; S. Y. Yue, Ma, and Long 2004; Ragonese, Scuderi, and Palmisano 2008; Chowdhury, Reynaert, and Niknejad 2009; Chan and Long 2010).

To give a simpler reading of the *TCR*, a direct relationship with the performance parameters of transformer coils is provided by the following simplified equation (5.6):

$$TCR \approx \omega \, Q_{EQ} L_{EQ} \qquad (5.6)$$

where Q_{EQ}, L_{EQ} are the equivalent Q-factor and inductance of the transformer seen at the primary coil, as defined in equations (5.7) and (5.8), respectively:

$$Q_{EQ} = Q_1 \frac{k^2 \, Q_1 \, Q_2}{1 + k^2 \, Q_1 \, Q_2} \cong Q_1 \qquad (5.7)$$

$$L_{EQ} = L_1\left(1 + \frac{1}{Q_1^2} + k^2\frac{Q_2}{Q_1}\right) \cong L_1\left(1 + k^2\frac{Q_2}{Q_1}\right) \tag{5.8}$$

Equation (5.6) clearly reveals the deep similarity between the TCR and the ωQL product of inductors and points out a strong dependence of the TCR on both the operating frequency and the coil inductance. These trends can be appreciated looking at Figures 5.7 and 5.8. Figure 5.7 reports the measured TCR values of stacked transformers with different metal widths (belonging to the set already used in Chapter 4) at four operating frequencies. As expected, the benefit of using a transformer tuned load boosts at higher operating frequencies. On the other hand, very small improvements are achievable only at low frequency by adopting wider spirals, thus indicating the relatively weak dependence of the TCR on the spiral Q-factors. Indeed, the most efficient procedure to improve the TCR performance at a given operating frequency is increasing the coil inductances by using multiturn spirals, as demonstrated in Figure 5.8. It is worth noting that the increment of n from 1.5 to 4.5 at fixed inner diameter and metal width (i.e., $w = 10$ μm, $d_{in} = 100$ μm) produces a reduction of the f_{SR} from 25 to 5.5 GHz. This remark suggests that the design of primary and secondary coils should account for the transformer self-resonance. For this reason, at very high operating frequencies the interleaved configuration is generally preferred to the stacked one because it is able to provide higher inductance values.

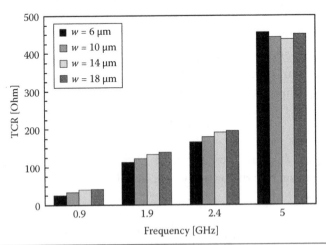

Figure 5.7 *TCR* of integrated stacked transformers as a function of the metal width at different operating frequencies ($n = 2.5$, $d_{in} = 100$ μm, $s = 4$ μm, $L_{DC1} = 1.2$ nH, $L_{DC2} = 1.3$ nH).

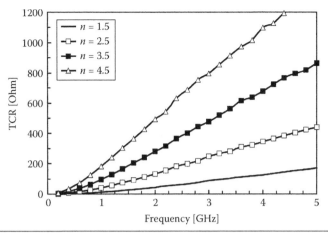

Figure 5.8   *TCR* of integrated stacked transformers as a function of the number of turns ($w =$ 10 μm, d_{in} = 100 μm, s = 4 μm)

5.4 Circuit Design Examples

To put into practice the design guidelines described above, three circuit examples are illustrated with the aim to cover different applications and main functional blocks of a RF/mm-wave transceiver front-end. The first two circuits—that is, an image-reject receiver front-end for 5-GHz WLANs and a 17-GHz voltage-controlled oscillator (VCO) —are implemented in a 45-GHz-f_T silicon bipolar process, already cited in the previous chapters, which provides three AlCu metal layers (3-μm-thick top metal), lateral pnp transistors, metal–insulator–metal (MIM) capacitors, and high-quality junction diode varactors (Biondi et al. 2003; Italia et al. 2005). In these two circuits, inductive components are electrically shielded from the underlying substrate via a buried layer radial PGS, which also guarantees a low-impedance path to the ground plane trough n⁺ sinker contacts, allowing a well-defined RF ground reference to be achieved.

The third example is a 24-GHz transmitter for automotive radar, which was implemented in a 0.13-μm SiGe BiCMOS technology featuring high-speed npn transistors with f_T/f_{max} of 166/175 GHz, dual V_T dual-gate oxide CMOS devices, and six-level metal copper back-end (Laurens et al. 2003). In this process both inductors and transformers take advantage of a polysilicon PGS to minimize the energy dissipated by the capacitive currents injected into the substrate.

5.4.1 An Image-Reject Receiver Front-End for 5-GHz WLANs

The proposed radio front-end was designed for a 5-GHz double-conversion WLAN receiver adopting a sliding-IF approach with the LOs running at $4f_{RF}/5$ and $f_{RF}/5$, respectively. The simplified schematic and the die micrograph of the receiver front-end are reported in Figure 5.9 and Figure 5.10, respectively. The circuit consists of a two-stage LNA

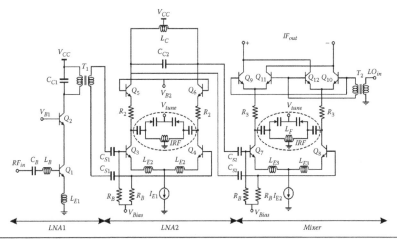

Figure 5.9 Simplified schematic of the 5-GHz receiver front-end with tunable IRFs. (From E. Ragonese, A. Scuderi, T. Biondi, and G. Palmisano "Scalable Lumped Modeling of Single-Ended and Differential Inductors for RF IC Design," *International Journal of RF and Microwave Computer-Aided Engineering*, vol. 19, pp. 110–119. © 2009 Wiley. With permission.)

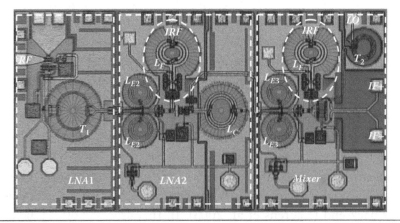

Figure 5.10 Die micrograph of the 5-GHz down-converter with tunable IRFs. (From E. Ragonese, A. Scuderi, T. Biondi, and G. Palmisano "Scalable Lumped Modeling of Single-Ended and Differential Inductors for RF IC Design," *International Journal of RF and Microwave Computer-Aided Engineering*, vol. 19, pp. 110–119. © 2009 Wiley. With permission.)

and a double-balanced mixer and includes two integrated image-reject filters (IRFs) to provide on-chip image-rejection ratio (IRR) higher than 50 dB. Both stages LNA1 and LNA2 adopt a cascode topology with resonant load, which allows high reverse isolation, power gain, and linearity to be achieved. The LNA1 was designed to achieve simultaneous minimum noise/input impedance matching, by using optimum transistor sizing and bonding wire base and emitter inductors (L_B, L_{E1}) (Girlando and Palmisano 1999). The stage is loaded by an integrated transformer (T_1), which implements both intermatching network (with capacitors C_{C1} and C_{S1}) and single-ended-to-differential conversion toward the LNA2 input. Both substrate and layout configurations of transformer T_1 were designed to maximize its TCR, which is related to the power gain (G_{P1}) of LNA1 according to equation (5.9):

$$G_{P1} \cong \frac{1}{2}\left(\frac{\omega_T}{2\omega_0}\right)^2 \frac{TCR}{\text{Re}\{Z_s\}} = \frac{1}{2}\left(\frac{\omega_T}{2\omega_0}\right)^2 \frac{(\omega_0 Q_{EQ} L_{EQ})}{\text{Re}\{Z_s\}} \tag{5.9}$$

where ω_T and ω_0 are the transition frequency of the transistor and the operating frequency, respectively; Z_S is the source impedance (i.e., 50 Ω); and Q_{EQ}, L_{EQ} are the equivalent Q-factor and inductance of the transformer seen at the primary coil, as defined in equations (5.7) and (5.8). The transformer T_1 features a symmetric layout, which is well suited to differential circuits and comprises two circular stacked coils, fabricated using the third and second metal layers. The stacked structure was chosen to achieve higher magnetic coupling and area efficiency compared to the interleaved counterpart. The geometric parameters of T_1 were properly set by taking advantage of EM simulations (i.e., $n = 2$, $w = 6$ μm, and $d_{in} = 220$ μm), and the closest spiral spacing allowed by the technology was adopted to exploit maximum magnetic coupling between adjacent metal paths.

Degeneration single-ended circular inductors (L_{E2} and L_{E3}) of around 2 nH were included in both LNA2 and the down-conversion mixer to improve the linearity performance. A differential circular inductor (L_C) of 2.7 nH was exploited as resonant load for the LNA2. It is worth mentioning that the design of the layout parameters of $L_{E2,3}$ ($n = 3$, $w = 7$ μm, $s = 3$ μm, $d_{in} = 98$ μm) and L_C ($n = 3$, $w = 12$ μm, $s = 3$ μm, $d_{in} = 140$ μm) followed the maximization procedures for Q and ωQL, respectively.

The tunable IRFs are included in both the LNA2 and mixer. They integrate differentially driven inductors (L_F), MIM capacitors, and octagonal junction diode varactors to implement a differential third-order filter, whose parallel and series resonance frequencies can be easily set to obtain an open circuit and a short circuit for the RF signal and the image signal, respectively (Ragonese, Italia, and Palmisano 2006), thus producing a notch in the power gain frequency response corresponding to the image band. To further improve the notch depth and hence the IRR, the current partitioning between the active circuits and the IRFs was optimized by means of resistances R_2 and R_3. Thanks to the varactor bias voltage V_{tune}, the notch frequency can be properly tuned, thus compensating for the fabrication tolerances. The use of a differential topology for the IRFs allows exploiting the benefits in terms of Q and f_{SR} of the differential configuration adopted for symmetric inductors L_F ($n = 3$, $w = 16$ μm, $s = 3$ μm, $d_{in} = 82$ μm).

The lumped scalable model presented in Chapter 3 was extensively adopted to set optimum spiral geometrical parameters, thus restricting EM simulations only to post-layout analysis (Ragonese et al. 2009a). The model was also exploited to choose the most advantageous inductor configuration (i.e., single-ended or differential), as trade-off between the device performance and area consumption.

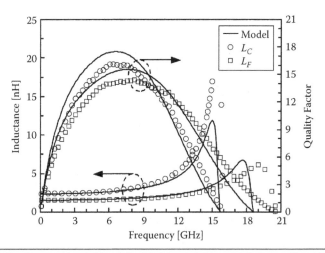

Figure 5.11 Measured (symbols) and simulated L and Q of L_C and L_F inductors. (From E. Ragonese, A. Scuderi, T. Biondi, and G. Palmisano "Scalable Lumped Modeling of Single-Ended and Differential Inductors for RF IC Design," *International Journal of RF and Microwave Computer-Aided Engineering*, vol. 19, pp. 110–119. © 2009 Wiley. With permission.)

As an example, Figure 5.11 reports modeled and measured L and Q for L_C and L_F inductors as a function of frequency. Maximum errors calculated at 5.3 GHz are lower than 3 and 10% for L and Q, respectively, thus confirming the soundness of the adopted design tool.

The receiver front-end was assembled in a leadless plastic package and characterized on a 400-μm-thick FR4 substrate. It achieves 19-dB power gain and a single-sideband noise figure of 4.8 dB, as shown in Figure 5.12, and the input 1-dB compression point is –23 dBm. The circuit draws 22 mA from the 3-V supply voltage. The IRR measurements are reported in Figure 5.13 as a function of three values of V_{tune}. At V_{tune} = 0.7 V, the receiver front-end attains a maximum IRR of 64 dB, while guaranteeing IRR better than 52 dB in the frequency range 5.15–5.55 GHz. Unlike other image filtering approaches reported in the literature (Lee, Samavati, and Rategh 2002; Rogers and Plett 2003; Nguyen et al. 2005), such a performance has been achieved without applying any Q-enhancement technique to the IRFs. Because the IRR specification for the adopted receiver architecture is around 60 dB, the presented front-end allows replacing the external high-quality IRF with a less selective low-priced RF filter.

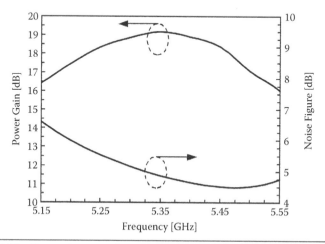

Figure 5.12 Measured power gain and noise figure of the 5-GHz down-converter. (From E. Ragonese, A. Scuderi, T. Biondi, and G. Palmisano "Scalable Lumped Modeling of Single-Ended and Differential Inductors for RF IC Design," *International Journal of RF and Microwave Computer-Aided Engineering*, vol. 19, pp. 110–119. © 2009 Wiley. With permission.)

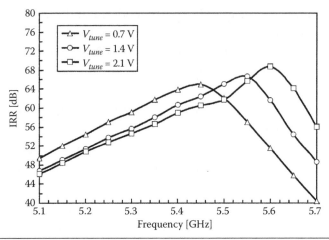

Figure 5.13   Measured IRR for different values of V_{tune}. (From E. Ragonese, A. Scuderi, T. Biondi, and G. Palmisano "Scalable Lumped Modeling of Single-Ended and Differential Inductors for RF IC Design," *International Journal of RF and Microwave Computer-Aided Engineering*, vol. 19, pp. 110–119. © 2009 Wiley. With permission.)

5.4.2   A 17-GHz Voltage-Controlled Oscillator with Transformer-Based Resonator

In recent years, the demand for VCOs featuring both low phase noise and wide tuning range promoted advanced solutions to improve the Q-factor of the resonator by optimizing both inductive components and varactors. In this design example, a transformer-based VCO for 17-GHz applications is presented (Scuderi and Palmisano 2006). To obtain an operative frequency of 17 GHz, the VCO adopts a 9-GHz core and a frequency doubler, consisting of a full-wave rectifier and a comparator. The simplified schematic and the die micrograph of the 17-GHz VCO are reported in Figures 5.14 and 5.15, respectively.

The 9-GHz VCO core takes advantage of an integrated transformer to implement a high-Q resonator and provide simultaneously the coupling between bases to collectors in the connection of the cross-coupled pair. The transformer is composed of three inductors L_V, L_C, and L_B connected to varactors, collectors, and bases, respectively. The adopted transformer-based VCO topology is able to overcome the main drawbacks of typical silicon bipolar LC VCOs (i.e., AM-PM

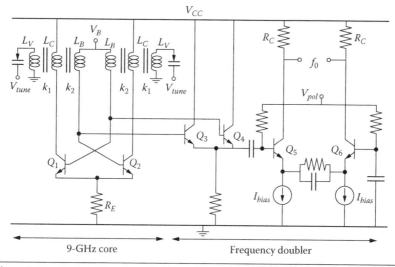

Figure 5.14   Simplified VCO schematic. (From A. Scuderi and G. Palmisano, 2006, "A Low-Phase-Noise Voltage-Controlled Oscillator for 17-GHz Applications," *IEEE Microwave and Wireless Components Letters,* vol. 16, pp. 191–193. © 2006 IEEE. With permission.)

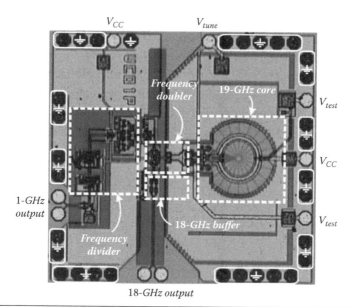

Figure 5.15   VCO die micrograph. (From A. Scuderi and G. Palmisano, 2006, "A Low-Phase-Noise Voltage-Controlled Oscillator for 17-GHz Applications," *IEEE Microwave and Wireless Components Letters,* vol. 16, pp. 191–193. © 2006 IEEE. With permission.)

noise conversion and reduced varactor capacitance variation $\Delta C_V/C_V$), which typically use the varactors connected to ground and to collectors by means of high-value resistors and coupling capacitors, respectively. Indeed, thanks to the inductive coupling between L_C and L_V, bias resistors and series capacitors can be removed from the resonator, thus allowing both low noise and maximum tuning range to be achieved. Moreover, the inductive coupling between L_C and L_B performs the connection between collectors and bases, thus avoiding coupling capacitors and biasing resistors in the feedback connection, as well.

The integrated transformer adopts three single-turn circular inter-leaved coils, which exploit the 3-μm-thick top metal available in the process. A close spacing (s = 3 μm) was adopted to maximize the magnetic coupling (k = 0.7) and different metal widths—that is, 10, 16, and 6 μm—were exploited for L_V, L_C, and L_B, respectively. The device has an outer diameter of 270 μm.

The LC resonator optimization is the key issue of the oscillator design. For these reasons, high-Q octagonal junction diode varactors with a wide tuning capability (around 80% with a voltage control sweep from 0 to 3 V) were used, and the transformer design was aimed at the minimization of the VCO phase noise. The geometrical parameters of the three interleaved coils were chosen by exploiting the lumped model for symmetric transformers shown in Figure 5.16 (Scuderi et al. 2006). The adopted transformer model was obtained as direct extension of the inductor model already presented in Chapter 3. Indeed, each single-turn spiral can be represented as stand-alone inductor, coupled to the others

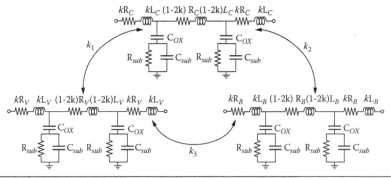

Figure 5.16 Model of a single-turn three-coil transformer. (From A. Scuderi et al., 2006, "A 18-GHz Silicon Bipolar VCO with Transformer-Based Resonator," *IEEE Radio Frequency Integrated Circuits Symposium Digest*, pp. 11–13. © 2006 IEEE. With permission.)

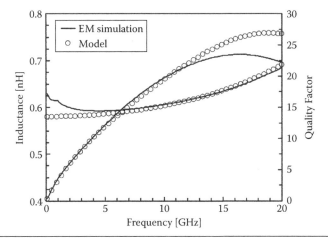

Figure 5.17   Simulated inductance and quality factor for L_v coil. (From A. Scuderi et al., 2006, "A 18-GHz Silicon Bipolar VCO with Transformer-Based Resonator," *IEEE Radio Frequency Integrated Circuits Symposium Digest*, pp. 11–13. © 2006 IEEE. With permission.)

through a proper coupling factor (k_i = 0.7). The accuracy of the transformer model was further confirmed by comparison with three-dimensional EM simulations, as reported in Figure 5.17. It is worth mentioning that the resonator design was refined by means of a post-layout EM simulation to take into account all connections and parasitics.

The transformer-based VCO was assembled in a leadless plastic package and characterized on a 400-μm-thick FR4 substrate at 3-V supply voltage. It exhibits phase noise of –109 dBc/Hz at 1-MHz offset from the center frequency of 18.5-GHz and 4.1-GHz tuning range (i.e., from 16.4 to 20.5 GHz), when the control voltage sweeps from 0 to 3 V. The core and the frequency doubler consume 8 and 4 mA, respectively. For a complete characterization of the circuit, the stand-alone 9-GHz core was integrated and tested (i.e., without frequency doubler) as well. The measured phase noise is –116 dBc/Hz at an offset frequency of 1 MHz from a carrier of 9.2 GHz. This result indicates that the measured phase noise at 18 GHz is slightly degraded by the frequency doubling circuitry with respect to the theoretical estimation (i.e., 6-dB phase noise increment).

A well-known figure of merit is usually adopted to evaluate the soundness of integrated VCOs by taking into account the power dissipation, the phase noise frequency dependency, as well as the tuning range (Tasic, Serdijn, and Long 2005). For the sake of completeness, the VCO figure of merit (FOM_{VCO}) expression is reported in equation (5.10):

$$FOM_{VCO} = L(\Delta f_{offset}) - 20\log\left(\frac{f_0}{\Delta f_{offset}}\right) + 10\log\left(\frac{P_{DC}}{1mW}\right) + 20\log\left(\frac{f_0}{\Delta f_{TUNE}}\right)$$

$$(5.10)$$

where $L(\Delta f_{offset})$ is the phase noise measured at the offset frequency Δf_{offset} from the carrier f_0, P_{DC} is the dc power consumption, and Δf_{TUNE} is the tuning range.

Experimental measurements indicate that the proposed 9- and 18-GHz oscillators achieve FOM_{VCO} as high as 169 and 166, respectively, thus outperforming state-of-the-art FOM_{VCO} values for X- and K-band oscillators implemented in f_T-equivalent technologies.

5.4.3 A 24-GHz UWB Transmitter for Automotive Short-Range Radar

This last design example concerns an emerging mm-wave application —that is, the automotive radar sensor—which is strategic for both public authorities and car suppliers to reduce the road accidents. Radar is universally identified as the eligible technology to face and solve this problem thanks to higher performance and robustness in comparison with other existing solutions. Radar systems allow obtaining a complete inspection around the car by means of long-range radar (LRR) and short-range radar (SRR) sensors. In particular, SRR sensors are able to detect a target in a range from 0.1 to 15–30 m, thus enabling a variety of applications concerning both comfort and safety fields, such as precrash sensing, collision mitigation, blind spot detection, parking aid, lane change assistant, rear crash collision warning, stop & go, and urban collision avoidance (Gresham et al. 2004). For this application the most important standardization authorities, such as the Federal Communication Commission (FCC) and the European Telecommunications Standards Institute (ETSI), allocated unlicensed bands around 24 GHz in the United States and European Union, respectively (FCC 2002; ETSI EN 302 288-1 2005).

However, due to the high cost of mm-wave electronics, traditionally addressed using III-V technologies, only high-end cars are equipped with 24-GHz radar sensors. In recent years, different silicon implementations (Krishnaswamy and Hashemi 2007; Mazzanti et al. 2008; Ragonese et al. 2009b) have proved the maturity of both high-speed

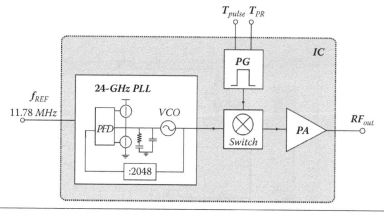

Figure 5.18 Block diagram of the 24-GHz UWB transmitter.

bipolar and sub-μm CMOS processes for 24-GHz automotive applications. Implementation of such systems using silicon-based processes can reduce the cost by an order of magnitude and increase the level of complexity, enabling ubiquitous and pervasive adoption of the automotive radar technology.

Figure 5.18 shows the simplified block diagram of the proposed 24-GHz UWB transmitter for automotive SRR sensors. The transmitter was designed in a 0.13-μm 166-GHz-f_T/175-GHz-f_{max} SiGe BiCMOS process and it is composed of an integrated 24-GHz frequency synthesizer implemented by means of a phase-locked loop (PLL), a switch driven by a pulse generator (PG), and a PA. The system is able to generate and transmit 24-GHz modulated pulses using different pulse widths (T_{pulse}) and pulse repetition interval (T_{PR}) in order to cover main sensor requirements for the SRR applications (Scuderi, Ragonese, and Palmisano 2009). In particular, these parameters settle on two important characteristics of a radar sensor, the resolution (i.e., the minimum distance between two targets that can be individually detected) and the maximum unambiguous range (i.e., the maximum distance at which a target can be detected). Typical values of T_{pulse} and T_{PR} in compliance with common SRR applications (e.g., parking and stop & go) are around 0.5–1 and 150–300 ns, respectively.

Figure 5.19 reports the simplified schematics of the high-frequency chain of the UWB transmitter, consisting of the 24-GHz VCO, the switch, and the PA. The VCO uses a bipolar core with a high-Q LC

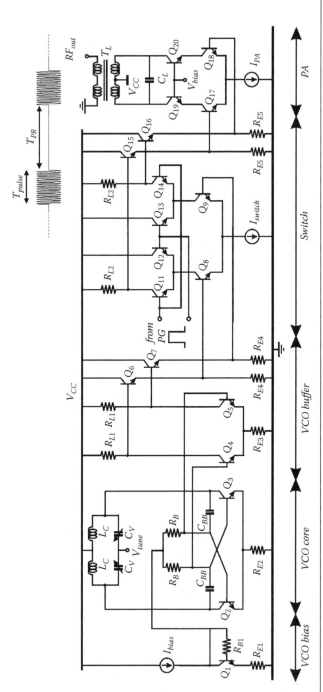

Figure 5.19 Simplified schematic of the 24-GHz UWB transmitter.

resonator. A buffer made up of a differential stage and two emitter follower stages is used at the oscillator output to drive the highly capacitive load exhibited by the switch (and the prescaler in the PLL). Both the VCO core and buffer form with Q_1 a current mirror, thus removing tail current sources. Indeed, a proper value of the emitter resistor R_{E2} allows the optimization of the oscillation amplitude and noise performance, while providing higher common-mode rejection and lower noise in comparison with traditional current sources. The VCO resonator exploits accumulation MOS variable capacitors, C_V, and a sub-nH spiral inductor L_C. L_C was implemented by means of single-turn octagonal coil and its structure was designed to reduce both series and substrate losses. Thanks to the advanced BEOL of the adopted process, it was possible to adopt a multilayer structure for the spiral inductors, consisting of two Cu metals (Metal 6 and Metal 5) plus a top aluminum layer. On the other hand, the substrate losses were highly reduced by means of a polysilicon conductive PGS, which does not appreciably affect the inductor f_{SR} thanks to the presence of a thick oxide layer beneath the Metal 5. The metal width of the spiral was carefully chosen to maximize the Q-factor ($w = 12$ μm). The EM simulations of L_C indicate L and Q of around 230 pH and 19 at 24 GHz, respectively.

The produced 24.125-GHz carrier is then pulsed by the sub-ns switch that consists of a differential current-steering stage driven by the PG. The PG is able to generate the control baseband pulse with different widths, from 1 to 0.5 ns, according to the required radar resolution (around 10 cm). Finally, the resulting 24-GHz UWB signal is amplified by the PA, which adopts a differential cascode topology with resonant load. The key component of the PA was the transformer T_L, which was designed to guarantee the required output power (up to 3 dBm) as well as differential-to-single-ended conversion of the RF signal. Moreover, its inherent galvanic isolation capability was exploited to increase robustness thus avoiding the losses of the electrostatic discharge protection structures. Load transformer T_L adopts an interleaved coil structure with a turn ratio of 2:1. The interleaved configuration allows a multilayer structure to be used for primary and secondary windings, thus maximizing the Q-factor for both coils. A magnetic coupling factor as high as 0.7 is achieved thanks to a tight metal spacing ($s = 2$ μm).

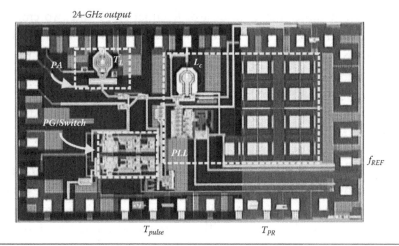

Figure 5.20   Die micrograph of the 24-GHz UWB transmitter.

The chip micrograph of the fabricated circuit is shown in Figure 5.20. Main blocks of the transmitter are highlighted and both inductor L_C and transformer T_L are labeled. It is worth noting that the adopted process, as common advanced BiCMOS and CMOS technologies, requires the use of active, polysilicon and metal dummies in compliance with well-defined density rules. It should be noticed that the generation of these dummies, and in particular of polysilicon and metal dummies, nearby inductive components is quite critical because conductive dummies can produce unpredictable and undesirable coupling effects, especially at high frequency. For this reason, no-dummy zones around L_C and T_L were adopted, in which the generation of conductive dummies was avoided.

The die was tested using a chip-on-board assembly technique on a 400-μm-thick FR4 substrate where supply filters were also mounted. This arrangement allows the high-frequency output to be directly measured on die using a GSG probe, thus reproducing a measurement setup comparable to the bumping flip-chip assembly (Heinrich 2005).

Figure 5.21 shows the measured power spectral density (PSD) at the transmitter (TX) output for a T_{pulse} and a T_{PR} of 1 and 270 ns, respectively. The two reported curves are obtained with and without the gain of the UWB patch antenna (not here described). The

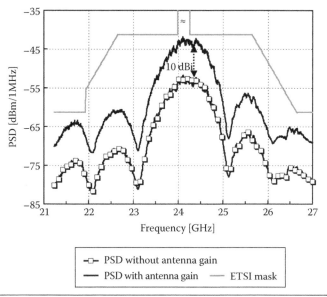

Figure 5.21   Measured PSD of the 24-GHz UWB transmitter ($T_{pulse} = 1$ ns, $T_{PR} = 270$ ns).

spectra present the main lobe centered at 24.125 GHz and two nulls at a 2-GHz span. By exploiting a 10-dBi antenna gain, the proposed transmitter complies with the transmission mask defined by the ETSI.

References

G. Avenier, M. Diop, P. Chevalier, et al., 2009, "0.13 μm SiGe BiCMOS Technology Fully Dedicated to mm-Wave Applications," *IEEE Journal of Solid-State Circuits*, vol. 44, pp. 2312–2321.

T. Biondi, F. Carrara, A. Scuderi, and G. Palmisano, 2003, "A Silicon Bipolar Technology for High-Efficiency Power Applications Up to C-Band," *IEEE Radio Frequency Integrated Circuits Symposium Digest*, pp. 155–158.

J. N. Burghartz, and B. Rejaei, 2003, "On the Design of RF Spiral Inductors on Silicon," *IEEE Transactions on Electron Devices*, vol. 50, pp. 718–729.

J. N. Burghartz, M. Soyuer, and K. Jenkins, 1996, "Integrated RF and Microwave Components in BiCMOS Technology," *IEEE Transactions on Electron Devices*, vol. 43, pp. 1559–1570.

F. Carrara, A. Italia, E. Ragonese, and G. Palmisano, 2006, "Design Methodology for the Optimization of Transformer Loaded RF Circuits," *IEEE Transactions on Circuits and Systems Part I*, vol. 53, pp. 761–768.

W. L. Chan, and J. R. Long, 2010, "A 58–65 GHz Neutralized CMOS Power Amplifier with PAE above 10% at 1-V Supply," *IEEE Journal of Solid-State Circuits*, vol. 45, pp. 554–564.

P. Chevalier, D. Gloria, P. Scheer, et al., 2006, "Advanced SiGe BiCMOS and CMOS Platforms for Optical and Millimeter-Wave Integrated Circuits," *Proceedings of the IEEE Compound Semiconductor Integrated Circuits Symposium*, pp. 12–15.

Y. S. Choi, and J. B. Yoon, 2004, "Experimental Analysis of the Effect of Metal Thickness on the Quality Factor in Integrated Spiral Inductors for RF ICs," *IEEE Electron Device Letters*, vol. 25, pp. 76–79.

D. Chowdhury, P. Reynaert, and A. M. Niknejad, 2009, "Design Considerations for 60 GHz Transformer-Coupled CMOS Power Amplifiers," *IEEE Journal of Solid-State Circuits*, vol. 44, pp. 2733–2744.

J. Craninckx, and M. S. J. Steyaert, 1997, "A 1.8-GHz Low-Phase-Noise CMOS VCO Using Optimized Hollow Spiral Inductors," *IEEE Journal of Solid-State Circuits*, vol. 32, pp. 736–744.

M. Danesh, and J. R. Long, 2002, "Differentially Driven Symmetric Microstrip Inductors," *IEEE Transactions on Microwave Theory and Techniques*, vol. 50, pp. 332–341.

T. O. Dickson, M.-A. LaCroix, S. Boret, D. Gloria, R. Beerkens, and S. P. Voinigescu, 2005, "30–100-GHz Inductors and Transformers for Millimeter-Wave (Bi)CMOS Integrated Circuits," *IEEE Transactions on Microwave Theory and Techniques*, vol. 53, pp. 123–133.

ETSI EN 302 288-1, 2005, *Electromagnetic Compatibility and Radio Spectrum Matters (ERM); Short Range Devices; Road Transport and Traffic Telematics (RTTT); Short Range Radar Equipment Operating in the 24 GHz Range; Part 1: Technical Requirements and Methods of Measurement.*

Federal Communication Commission, 2002, *First Report and Order, Revision of Part 15 of the Commission's Rules Regarding Ultra Wideband Transmission Systems*, FCC, Washington, DC. ET Docket 98 153.

G. Girlando, and G. Palmisano, 1999, "Noise Figure and Impedance Matching in RF Cascode Amplifiers," *IEEE Transactions on Circuits and Systems Part II*, vol. 46, pp. 1388–1396.

G. Girlando, E. Ragonese, and G. Palmisano, 2004, "Silicon Bipolar LNA's in the X and Ku Bands," *Analog Integrated Circuits and Signal Processing*, vol. 41, pp. 119–127.

I. Gresham, A. Jenkins, R. Egri, et al., 2004, "Ultra-Wideband Radar Sensors for Short-Range Vehicular Applications," *IEEE Transactions on Microwave Theory Techniques*, vol. 52, pp. 2105–2122.

W. Heinrich, 2005, "Approach for Millimeter-Wave Packaging," *IEEE Microwave Magazine*, vol. 6, pp. 35–45.

A. Italia, F. Carrara, E. Ragonese, T. Biondi, A. Scuderi, and G. Palmisano, 2005, "The Transformer Characteristic Resistance and Its Application to the Performance Analysis of Silicon Integrated Transformers," *IEEE Radio Frequency Integrated Circuits Symposium Digest*, pp. 597–600.

A. Italia, L. La Paglia, A. Scuderi, F. Carrara, E. Ragonese, and G. Palmisano, 2005, "A Silicon Bipolar Transmitter Front-End for 802.11a and HIPERLAN2 Wireless LANs," *IEEE Journal of Solid-State Circuits*, vol. 40, pp. 1451–1459.

M. Khanpour, K. W. Tang, P. Garcia, and S. P. Voinigescu, 2008, "A Wideband W-Band Receiver Front-End in 65-nm CMOS," *IEEE Journal of Solid State Circuits*, vol. 43, pp. 1717–1730.

Y. K. Koutsoyannopoulos, and Y. Papananos, 2000, "Systematic Analysis and Modeling of Integrated Inductors and Transformer in RF IC Design," *IEEE Transactions on on Circuits and Systems Part II*, vol. 47, pp. 699–712.

H. Krishnaswamy, and H. Hashemi, 2007, "A Fully Integrated 24GHz 4-Channel Phased-Array Transceiver in 0.13µm CMOS Based on a Variable-Phase Ring Oscillator and PLL Architecture," *International Solid State Circuits Conference Digest Technical Papers*, pp. 124–125.

E. Laskin, P. Chevalier, A. Chantre, B. Sautreuil, and S. P. Voinigescu, 2008, "165-GHz Transceiver in SiGe Technology," *IEEE Journal of Solid-State Circuits*, vol. 43, pp. 1087–1100.

M. Laurens, B, Martinet, O. Kermarrec, et al., 2003, "A 150 GHz f_T/f_{max} 0.13 µm SiGe:C BiCMOS Technology," *Proceedings of the IEEE Bipolar/BiCMOS Circuits Technology Meeting*, pp. 199–202.

T. H. Lee, H. Samavati, and H. R. Rategh, 2002, "A 5-GHz CMOS Wireless LANs," *IEEE Transactions on Microwave Theory and Techniques*, vol. 50, pp. 268–280.

C.-C. Lim, K.-S. Yeo, K.-W. Chew, et al., 2008, "High Self-Resonant and Area Efficient Monolithic Transformer Using Novel Intercoil-Crossing Structure for Silicon RFIC," *IEEE Electron Device Letters*, vol. 29, pp. 1376–1379.

Y.-S. Lin, 2005, "Implementation of Perfect-Magnetic-Coupling Ultralow-Loss Transformer in RFCMOS Technology," *IEEE Electron Device Letters*, vol. 26, pp. 832–835.

J. R. Long, 2000, "Monolithic Transformers for Silicon RF IC Design," *IEEE Journal of Solid-State Circuits*, vol. 35, pp. 1368–1382.

J. R. Long, and M. A. Copeland, 1997, "The Modeling, Characterization, and Design of Monolithic Inductors for Silicon RF IC's," *IEEE Journal of Solid-State Circuits*, vol. 32, pp. 357–369.

A. Mazzanti, M. Sosio, M. Repossi, and F. Svelto, 2008, "A 24GHz Sub-Harmonic Receiver Front-End with Integrated Multi-Phase LO Generation in 65nm CMOS," *International Solid State Circuits Conference Digest Technical Papers*, pp. 216–217.

S. S. Mohan, M. del Mar Hershenson, S. P. Boyd, and T. H. Lee, 1999, "Simple Accurate Expressions for Planar Spiral Inductances," *IEEE Journal of Solid-State Circuits*, vol. 34, pp. 1419–1424.

O. H. Murphy, K. G. McCarthy, C. J. P. Delabie, A. C. Murphy, and P. J. Murphy, 2005, "Design of Multiple-Metal Stacked Inductors Incorporating an Extended Physical Model," *IEEE Transactions on Microwave Theory and Techniques*, vol. 53, pp. 2063–2072.

K. T. Ng, B. Rejaei, and J. N. Burghartz, 2002, "Substrate Effects in Monolithic RF Transformers on Silicon," *IEEE Transactions on Microwave Theory and Techniques*, vol. 50, pp. 377–383.

T.-K. Nguyen, N.-J. Oh, C.-Y. Cha, et al., 2005, "Image-Rejection CMOS Low-Noise Amplifier Design Optimization Techniques," *IEEE Transactions on Microwave Theory and Techniques*, vol. 53, pp. 538–547.

A. Parsa, and B. Razavi, 2009, "A New Transceiver Architecture for the 60-GHz Band," *IEEE Journal of Solid-State Circuits*, vol. 44, pp. 751–762.

C. Pastore, F. Gianesello, D. Gloria, J. C. Giraudin, O. Noblanc, and P. Benech, 2009, "High Performance and High Current Integrated Inductors Using a Double Ultra Thick Copper Module in an Advanced 65 nm RF CMOS Technology," *IEEE Topical Meeting on Silicon Monolithic Integrated Circuits in RF Systems Technical Digest*, pp. 1–4.

E. Ragonese, A. Italia, and G. Palmisano, 2006, "An Image Reject Down Converter for 802.11a and HIPERLAN2 Wireless LANs," *Springer Telecommunication Systems*, vol. 32, pp. 105–115.

E. Ragonese, A. Scuderi, T. Biondi, and G. Palmisano, 2009, "Scalable Lumped Modeling of Single-Ended and Differential Inductors for RF IC Design," Wiley, *International Journal of RF and Microwave Computer-Aided Engineering*, vol. 19, pp. 110–119.

E. Ragonese, A. Scuderi, V. Giammello, et al., 2009, "A Fully Integrated 24GHz UWB Radar Sensor for Automotive Applications," *International Solid State Circuits Conference Digest Technical Papers*, pp. 306–307.

E. Ragonese, A. Scuderi, and G. Palmisano, 2008, "A Transformer-Loaded Variable-Gain LNA for 24-GHz Vehicular Short-Range Radar," *Microwave and Optical Technology Letters*, vol. 50, pp. 2013–2016.

B. Razavi, 1998, *RF Microelectronics*, Prentice-Hall, Englewood Cliffs, NJ.

J. W. M. Rogers, and C. Plett, 2003, "A 5-GHz Radio Front-End with Automatically Q-Tuned Notch Filter and VCO," *IEEE Journal of Solid-State Circuits*, vol. 38, pp. 1547–1554.

F. Rotella, B. K. Bhattacharya, V. Blaschke, et al., 2005, "A Broad-Band Lumped Element Analytic Model Incorporating Skin Effect and Substrate Loss for Inductors and Inductor Like Components for Silicon Technology Performance Assessment and RFIC Design," *IEEE Transactions on Electron Devices*, vol. 52, pp. 1429–1441.

F. M. Rotella, C. Cismaru, Y. Tkachenko, Y. Cheng, and P. J. Zampardi, 2006, "Characterization, Design, Modeling, and Model Validation of Silicon-Wafer M:N Balun Components under Matched and Unmatched Conditions," *IEEE Journal of Solid-State Circuits*, vol. 41, pp. 1201–1209.

A. Scuderi, T. Biondi, E. Ragonese, and G. Palmisano, 2005, "Analysis and Modeling of Thick Metal Spiral Inductors on Silicon," *Proceedings of the European Microwave Conference*, vol. 1, pp. 81–84.

A. Scuderi, and G. Palmisano, 2006, "A Low-Phase-Noise Voltage-Controlled Oscillator for 17-GHz Applications," *IEEE Microwave and Wireless Components Letters*, vol. 16, pp. 191–193.

A. Scuderi, E. Ragonese, T. Biondi, and G. Palmisano, 2006, "A 18-GHz Silicon Bipolar VCO with Transformer-Based Resonator," *IEEE Radio Frequency Integrated Circuits Symposium Digest*, pp. 11–13.

A. Scuderi, E. Ragonese, and G. Palmisano, 2009, "24-GHz Ultra-Wideband Transmitter for Vehicular Short-Range Radar Applications," *IET Circuits, Devices and Systems*, vol. 3, pp. 313–321.

L. Shiwei, and G. Lihui, 2003, "Influence of Metal Layer Thickness of Spiral Inductors on the Quality Factor by 3-D EM Simulation," *Proceedings of the 5th International Conference on ASIC*, vol. 2, pp. 1117–1119.

C. B. Sia, B. H. Ong, K. W. Chan, et al., 2005, "Physical Layout Design Optimization of Integrated Spiral Inductors for Silicon-Based RFIC Applications," *IEEE Transactions on Electron Devices*, vol. 52, pp. 2559–2567.

W. Simbürger, H.-D. Wohlmuth, P. Weger, and A. Heinzm, 1999, "A Monolithic Transformer Coupled 5-W Silicon Power Amplifier with 59% PAE at 0.9 GHz," *IEEE Journal of Solid-State Circuits*, vol. 34, pp. 1881–1892.

A. Tasic, W. A. Serdijn, and J. R. Long, 2005, "Design of Multistandard Adaptive Voltage-Controlled Oscillators," *IEEE Transactions on Microwave Theory and Techniques*, vol. 53, pp. 556–563.

C. P. Yue, and S. S. Wong, 1998, "On-Chip Spiral Inductors with Patterned Ground Shields for Si-Based RF ICs," *IEEE Journal of Solid-State Circuits*, vol. 33, pp. 743–752.

S. Y. Yue, D. Ma, and J. R. Long, 2004, "A 17.1–17.3-GHz Image-Reject Downconverter with Phase-Tunable LO Using 3× Subharmonic Injection Locking," *IEEE Journal of Solid-State Circuits*, vol. 39, pp. 2321–2332.

6

INDUCTIVE COMPONENTS ON
DIELETRIC SUBSTRATES

The current trend to fabricate modern wireless systems featuring high-frequency performance and high-speed digital capability, such as third-generation mobile cellular phones, takes advantage of the MCM approach. It is an advanced assembly technology that allows high-performance RF/mm-wave ICs, highly integrated digital ICs, and low-cost compact passive components such as filters, couplers, balanced-to-unbalanced transformers (baluns), diplexer, etc., to be packaged together. This solution represents a flexible and cost-effective alternative to the fully integrated approach because it allows individual parts of a complex system to be fabricated separately using the most suitable technology. The availability of high-quality integrated passive devices (IPDs) fabricated on insulating substrates, such as glass, is a key factor that enables cost-competitive solutions for such systems.

Another promising technology that might create innovative applications and open new, widespread, and diversified market segments is large area electronics on flexible substrates (Jain et al. 2005). Their flexibility, low-cost, high-volume manufacturing, excellent biocompatibility, and capability of covering large surfaces allow invisible embedding of electronics in several consumer goods (e.g., packaging, clothing, etc.), give new opportunities to automotive and architectural glazing industries, boost the development of smart systems for biomedical applications, and open a road to throw-away electronics. High-quality inductive components fabricated on plastic substrates are considered fundamental bricks to develop flexible RF ICs (Ma and Su 2009).

In this chapter, the fabrication of passive devices on both rigid (i.e., glass) and flexible (i.e., plastic) insulating substrates is discussed. Advantages and drawbacks of such technologies will be highlighted and compared with their silicon counterpart. After a quick overview of the state of the art of IPD technologies reported in Section 6.1, the design of a MCM for 5-GHz WLAN with passive components fabricated on glass is described in Section 6.2. Passive devices on plastic substrates are discussed in Section 6.3, where experimental measurements and lumped scalable modeling of spiral inductors are reported.

6.1   IPD Technology

The inherent advantage of IPD technologies relies on the easiness of fabricating passive devices with excellent performance in the RF/mm-wave frequency range, still keeping low cost and small size. For this reason, several foundries have developed IPD technologies on silicon, GaAs, or glass substrates. One or two (thick) copper metal layers could be available to allow routing of inductors and transformers with low dc resistance. In Kim et al. (2009) an IPD technology on silicon substrate is used to design a high-performance power divider at 5 GHz. This technology features two aluminum metal levels for component connections and capacitor plats and one 8-μm-thick copper metal layer to build spiral inductors. Thanks to such a high thickness inductors feature quality factor at 5 GHz higher than 70. The technology allows flip-chip design by using under-bump metal deposition.

In Huang et al. (2008) a Marchand balun is designed and fabricated using an IPD technology on glass substrate. Also in this case, the process provides three metal layers with a 10-μm-thick copper top metal layer for inductive components.

In Chen et al. (2009) the design of a CMOS LNA with IPD components is presented. Thanks to IPD components and CMOS amplifier codesign, the LNA exhibits state-of-the-art performance in terms of bandwidth, gain, and noise figure.

The 5-GHz MCM discussed in Section 6.2 uses an IPD process on glass (Bonnet, Dupont, and Berens 2007), which provides two thin AlCu metal layers for capacitor plats and component connection, two thick Cu metal layers for high-quality passive components (e.g., inductors, transformers, microstrips, and coplanar waveguides), and under-bump metallization for flip-chip assembly.

Since 2005 several examples of IPDs on plastic substrates were published (Guo, Zhang, Li, et al. 2005; Guo, Zhang, Lo, et al. 2005;

Kao et al. 2005; Teh et al. 2005), reporting Q-factors higher than 30 for spiral inductors with typical inductance values used for RF ICs.

In Ravesi et al. (2008), an IPD technology on polyethylene naphthalate (PEN) is presented. The process features Au metals for inductive components, MIM capacitors, and TaN resistors of 35–200 ohm/ square. Section 6.3 reports an extensive performance analysis of spiral inductors fabricated in this technology platform.

Figure 6.1 displays the Q-factor of three 2.5-turn, 20-μm-wide spiral inductors of around 1.8 nH fabricated on glass, plastic, and silicon substrates. The comparison highlights the most important advantages of inductors fabricated on glass or plastic with respect to their silicon counterpart; that is, reduced substrate losses thanks to negligible magnetically induced eddy currents and very low vertical displacement currents and reduced capacitive parasitics. These advantages allow a considerable enhancement of both the self-resonance frequency and the Q-factor, which is mainly limited by series resistive losses (i.e., skin and proximity effects). As shown in Figure 6.1, glass and plastic exhibit peak Q-factors as high as 50 and 40, respectively, whereas silicon substrate limits Q below 15. Moreover, dielectric substrates allow operation over a much higher frequency range, because their f_{SR} is more than doubled with respect to silicon. Although inductors fabricated on glass and plastic substrates show comparable performance, the latter solution allows a significant reduction

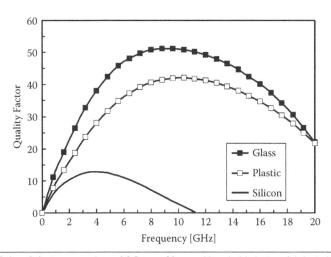

Figure 6.1 Q-factor comparison of 2.5-turn, 20-μm-wide spiral inductors fabricated on glass, plastic, and silicon substrates.

of costs thanks to cheaper manufacturing processes, for example, sheet-to-sheet or roll-to-roll processes (Jain et al. 2005).

6.2 Inductive Components on Glass Substrate

The advantages of high-Q passive devices in IPD technology are now discussed by describing the design of a 5-GHz WLAN module prototype composed of a silicon RF front-end and copper passive devices fabricated on glass substrate. Starting from state-of-the-art solutions (Zargari et al. 2002; Behzad et al. 2003; Italia et al. 2004), a WLAN front-end architecture and its specification were defined to comply with ETSI HIPERLAN2 and IEEE 802.11a standard constraints. A sliding-IF superheterodyne RF front-end, including down-converter, up-converter, and frequency synthesizer, was designed and fabricated in silicon technology (Italia et al. 2008). The front-end was designed to be mounted by flip-chip on a glass module fabricated in an IPD technology, whose passive devices were exploited for the design of both the image rejection and transmitter filters. The RF module was assembled in a low-cost plastic package.

6.2.1 5-GHz WLAN Module Architecture

The silicon RF front-end is based on a superheterodyne architecture with a sliding IF of 1 GHz (Zargari et al. 2002). Sliding IF with respect to traditional superheterodyne architectures uses only a single-frequency synthesizer. According to the frequency plan detailed in Table 6.1, the incoming RF signal at 5 GHz is down-converted to 1-GHz intermediate frequency (f_{IF}) by using a local oscillator LO_1 at 4 GHz. Then, the f_{IF} is down-converted to baseband signal by using $LO_2 = LO_1/4$. In spite of the easiness of the frequency plan, this system requires two key components: an IRF to suppress the image

Table 6.1 Frequency Plan

	FREQUENCY (GHz)
f_{RF}	5.18–5.70
$f_{LO1} = (4/5) \cdot f_{RF}$	4.14–4.56
$f_{IM} = f_{LO1} - f_{IF}$	3.11–3.42
$f_{IF} = f_{LO2} = (1/5) \cdot f_{RF}$	1.03–1.14

frequency (f_{IM}) and a TX filter to suppress the LO frequency (f_{LO}). From a system-level analysis, IRFs and TX filters require rejections of 60 and 40 dBc at f_{IF} and f_{LO}, respectively.

The block diagram of the module is sketched Figure 6.2, which highlights the silicon RF front-end in the gray box, along with the passive components designed in IPD technology. All off-chip components—that is, two highly selective filters, baluns, supply filtering, and connections—are fabricated in IPD technology. This approach avoids the inherent problems of direct conversion, allowing low-cost off-chip components. Both IRFs and TX filters on the WLAN module were designed by using the S-parameter measurements of three stacked inductors (labeled L_1, L_2, and L_3), whose geometrical details and measured performance in terms of inductance and Q-factor at 5 GHz are reported in Table 6.2.

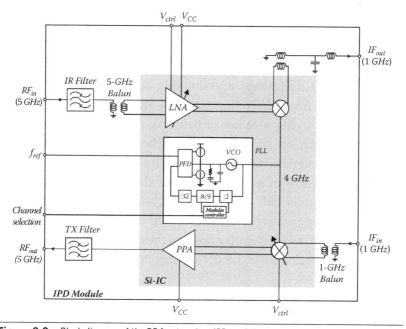

Figure 6.2 Block diagram of the RF front-end on IPD module.

Table 6.2 Inductor Geometrical Parameters and Measured Performance at 5 GHz

INDUCTOR	w (μm)	d_{in} (μm)	n	L (nH)	Q
L_1	50	300	1	1.5	75
L_2	90	200	1	1.1	85
L_3	90	400	1	2.2	75

The silicon IC is composed of the down-converter, up-converter, and frequency synthesizer. The down-converter consists of a variable-gain LNA and a double-balanced mixer. The LNA makes use of a fully differential configuration and is designed to achieve simultaneous noise/input matching by using optimum transistor size, on-chip inductive emitter degeneration, and input-matching inductors fabricated in IPD. The load of the down-converter mixer is an IPD transformer as well. The up-converter is composed of a variable-gain double-balanced mixer and a prepower amplifier. A 1-GHz IPD balun is used at the input of the mixer to provide the single-ended-to-differential conversion. The frequency synthesizer generates a LO_1 in the 4–4.6 GHz range and the programmable divider allows channel selection.

6.2.2 IRFs and TX Filters in IPD Technology

The simplified block diagram of the down-converter is shown in Figure 6.3. HIPERLAN2 and IEEE 802.11a specifications require 90 dBc image rejection for the overall system. Consequently, taking into account 30-dBc image rejection due to antenna selectivity, the receiver section (i.e., LNA, mixer, and IR filter) must provide a rejection as high as 60 dBc with low in-band loss to preserve the RX chain noise figure. To reject the frequency band between 3.1 and 3.4 GHz, the IRF was designed according to a notch topology.

The IRF features an in-band loss lower than 1.3 dB and an image rejection as high as 80 dBc, as highlighted in Figure 6.4(a) and Figure 6.4(b), respectively. Worst cases are also reported considering capacitor tolerance of ±5% with respect to the typical value. The in-band loss has a worst value about 1.5 dB at the minimum in-band frequency

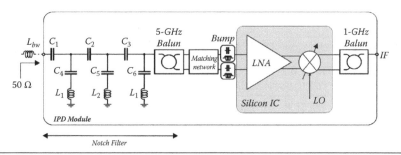

Figure 6.3 Block diagram of the down-converter chain on IPD.

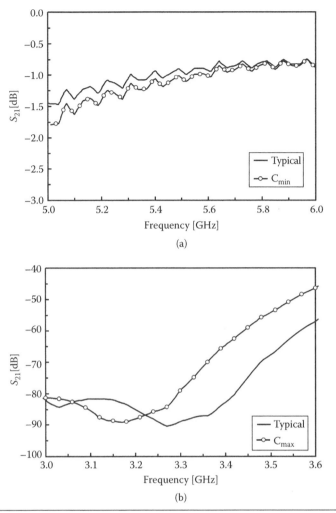

Figure 6.4 Typical and worst-case (C_{min}) in-band (5.18–5.7 GHz) loss of the IRF (a) and typical and worst-case (C_{max}) rejection of the image-frequency band (3.11–3.42 GHz) (b).

for minimum capacitor corner process (C_{min} case). In this case the filter stop-band presents a slight variation (not reported). On the contrary, when a process with maximum capacitor values (C_{max}) occurs, the image rejection decreases up to 62 dBc in the upper image frequency.

The receiver input network also provides a single-ended-to-differential conversion by using a 5-GHz balun with a measured insertion loss as low as 0.4 dB. A matching network between the balun and the input of the LNA was designed to provide both noise and input matching optimization.

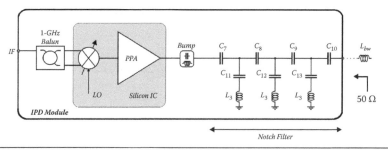

Figure 6.5 Block diagram of the up-converter chain on IPD.

The block diagram of the up-converter is shown in Figure 6.5. The IPD transmitter section is composed of an output TX filter based on a notch topology and an input 1-GHz balun for single-ended-to-differential conversion. The transmission dynamic range depends on the LO-to-carrier suppression and consequently the system analysis suggests a LO rejection as high as 40 dBc. Unfortunately, because the LO frequency band is very close to the RF band, such LO rejection can be achieved at expense of a high-sensitivity in-band loss. The TX filter characteristic is shown in Figure 6.6. The filter provides an in-band loss lower than 2 dB and an image rejection as high as 45 dBc, as highlighted in Figure 6.6(a) and Figure 6.6(b), respectively. Worst cases are also reported considering capacitor tolerance of ±5% with respect to the typical value. For minimum capacitor values (C_{min} case) the in-band loss exhibits a worst value about 3 dB at the minimum in-band frequency. In the case of a process with maximum capacitor values (C_{max}) the image rejection decreases up to 40 dBc for the upper LO frequency.

In the design of IPD networks the assembly strategy plays a crucial role. In this case the module is assembled in a low-cost plastic package and the RF silicon is connected to the IPD module by using flip-chip technology. Inductances of bumps for flip-chip connections and package bond wires are estimated by using three-dimensional EM simulation and taken into account in the design of IPD components. Input and output return loss of stand-alone TX and RX networks are lower than –18 dB at the operative frequency. Unfortunately, a bonding wire inductance L_{bw} at the input of the module produces some mismatch. Simulations indicate that a maximum L_{bw} value of 1 nH allows obtaining S_{11} lower than –15 dB. To obtain such condition, multiple bonding wires are used at the input of the module.

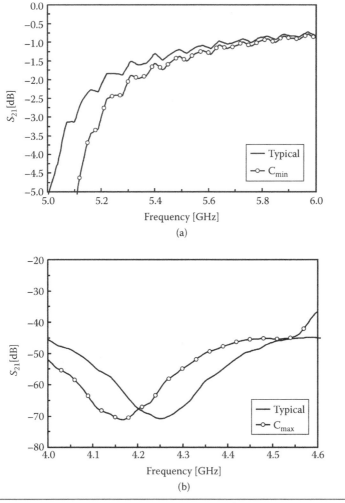

Figure 6.6 Typical and worst-case (C_{min}) in-band (5.18–5.7 GHz) loss of the TX filter (a) and typical and worst-case (C_{max}) suppression of the LO frequency (4.14–4.56 GHz) (b).

The layout of the module along with the IC micrograph is shown in Figure 6.7. The module size is 5.5 × 5.5 mm² and it was mounted in a 36-lead 8 × 8 mm² plastic package. The silicon die occupies 2.2 × 2.2 mm². A large number of bumps were exploited to ensure an excellent contact between silicon IC and the IPD ground planes. Both in the module and in the silicon chip, three different ground planes were used for the down-converter, the up-converter, and the frequency synthesizer to minimize cross-coupling. A large number of down-bonding

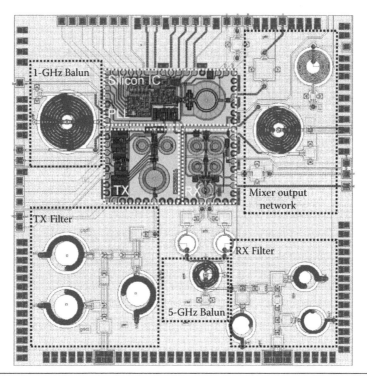

Figure 6.7   IPD module layout with IC micrograph.

connections were also used to minimize the parasitic inductances and improve the ground reference.

Thanks to the codesign between the integrated RF front-end and IPD components on the module, the prototype features an image-rejection ratio as high as 88 dBc with an overall down-converter noise figure as low as 4.5 dB. In the transmission section, the IPD output filter allows an output power of 0 dBm (i.e., an error vector magnitude of –34 dB) with a dynamic range higher than 35 dB.

6.3   Inductive Components on Plastic Substrate

A wide set of geometrically scaled circular spiral inductors has been designed and fabricated in the IPD process on PEN described in (Ravesi et al. 2008), with the aim of evaluating both performance and geometrical dependencies within the typical inductance range exploited by RF ICs. The inductors used 3- and 1-μm-thick Au metal layers for the spiral and the underpass, respectively, and had metal

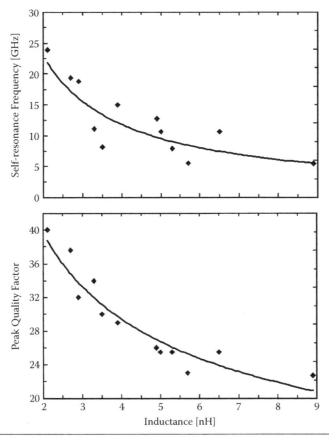

Figure 6.8 Measured self-resonance frequency and Q-factor performance of spiral inductors on PEN as function of inductance ($s = 10$ μm).

width (w) from 20 to 60 μm, inner diameter (d_{in}) from 100 to 250 μm, and turn number (n) from 2.5 to 4.5. The intermetal spacings (s) were set to 10 and 20 μm to comply with typical resolutions of wafer-to-wafer (W2W) and roll-to-roll (R2R) processing modes, respectively. Inductor performance is demonstrated in Figure 6.8 and Figure 6.9, which report measured f_{SR} and peak Q as a function of the low-frequency inductance (L_{DC}) for 10- and 20-μm-spaced spirals, respectively. As expected, higher Q-factors are achieved by exploiting the closest spacing (i.e., 10 μm), thanks to the higher magnetic coupling between the spiral turns. On the other hand, an increased spacing can produce some advantage in terms of f_{SR} at high inductance values due to the reduced fringing capacitance. Indeed, it is worth noting that capacitive parasitics toward the plastic substrate are

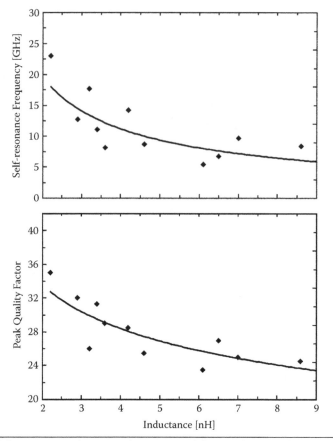

Figure 6.9 Measured self-resonance frequency and Q-factor performance of spiral inductors on PEN as function of inductance ($s = 20$ μm).

practically negligible and main contributions to self-resonance derive from the fringing capacitance and especially from the capacitance between the spiral and the underpass, which are separated by an oxide layer of around 1 μm.

The lack of significant substrate losses suggests focusing the attention on the reduction of the series losses in the coil. Obviously, thick metals give a considerable technology advantage (especially at low frequencies), and the designer can improve inductor performance by using a proper spiral width. Figure 6.10 depicts the Q-factor measurements of three 3.5-turn spirals at increasing metal widths, from 20 to 60 μm. The inductors, which have similar L_{DC} values (i.e., 5, 5.3, and 5.7 nH for w of 20, 40, and 60 μm, respectively), exhibit different frequency behaviors due to different values of the f_{SR} (i.e., 5.5, 8, and 12.5 GHz

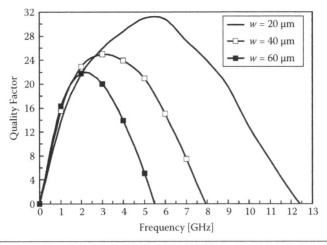

Figure 6.10 *Q*-factor of spiral inductors on PEN as a function of the metal width ($n = 3.5$, $d_{in} = 200$ μm, $s = 10$ μm).

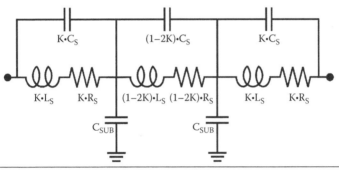

Figure 6.11 Proposed model for spiral inductors on PEN.

for w of 60, 40, and 20 μm, respectively). Below 3 GHz the benefit of wide metals can be appreciated thanks to the slight increase in the *Q*-curve slope. However, this advantage is fruitless at higher frequencies due to a significant reduction of the f_{SR}, which is mainly ascribed to the increment of the parasitic capacitance between the spiral and the underpass. This comparison confirms that the optimization of the metal width is highly recommended, as already explained in Chapter 5 for Si-integrated inductors.

A simple lumped model for spiral inductors on PEN was also developed by taking advantage of measured data of 10-μm-spaced spirals. The model is sketched in Figure 6.11 and it clearly derives from the one reported in Chapter 3 for spiral inductors on silicon. Indeed, the model adopts the splitting factor K (whose value is set from 0.1 to 0.2 according to the f_{SR}) to obtain a better approximation of a distributed network. Inductance L_S is calculated by means of equation (3.4). C_S mainly accounts for the capacitance between the spiral and the underpass, and the parasitic capacitance toward the substrate (C_{SUB}) is reported for the sake of completeness, although its effect is negligible. To obtain high accuracy in the estimation of the Q-factor, the model exploits a frequency-dependent expression for the series resistance R_S, which was derived from equation (3.8) using fitting parameters to accounts for both skin (proportional to f) and proximity (proportional

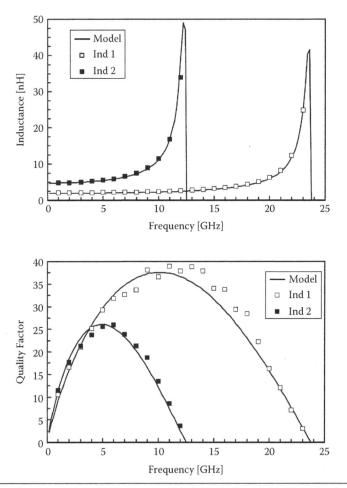

Figure 6.12 Comparisons between measured (symbols) and simulated (solid lines) inductances and Q-factors for spiral inductors on PEN (Ind 1: $n = 2.5$, $w = 20$ μm, $d_{in} = 150$ μm, $s = 10$ μm; Ind 2: $n = 3.5$, $w = 20$ μm, $d_{in} = 100$ μm, $s = 10$ μm).

to f^2) effects. Within the above-mentioned ranges for the geometrical parameters, maximum errors of Q are below 10%. As an example of model prediction capability, the comparisons between measured and simulated inductances and Q-factors of two spiral inductors on PEN (Ind 1: $n = 2.5$, $w = 20$ μm, $d_{in} = 150$ μm, $s = 10$ μm; Ind 2: $n = 3.5$, $w = 20$ μm, $d_{in} = 100$ μm, $s = 10$ μm) are reported in Figure 6.12.

References

A. R. Behzad, M. S. Zhong, S. B. Anand, et al., 2003, "A 5-GHz Direct-Conversion CMOS Transceiver Utilizing Automatic Frequency Control for the IEEE 802.11a Wireless LAN Standard," *IEEE Journal of Solid-State Circuits*, vol. 38, pp. 2209–2220.

B. Bonnet, F. Dupont, and F. Berens, 2007, "Ultra Wide Band Miniature Antenna," *Proceedings of IEEE Conference on Ultra-Wideband*, pp. 678–682.

H.-K. Chen, Y.-C. Hsu, T.-Y. Lin, et al., 2009, "CMOS Wideband LNA Design Using Integrated Passive Device," *IEEE Microwave Symposium Digest*, pp. 673–676.

L. H. Guo, Q. X. Zhang, H. Y. Li, et al., 2005, "Successful Transferring of Active Transistors, RF-Passive Components and High Density Interconnect from Bulk Si to Organic Substrates," *Technical Digest - International Electron Devices Meeting*, pp. 675–678.

L. H. Guo, Q. X. Zhang, G. Q. Lo, N. Balasubramanian, and D.-L. Kwong, 2005, "High-Performance Inductors on Plastic Substrate," *IEEE Electron Device Letters*, vol. 26, pp. 619–621.

H.-C. Huang, T.-C. Wei, T.-S. Horng, et al., 2008, "Design and Implementation of Spiral-Type Marchand Balun Using Glass-Based IPD Technology," *Proceedings of the IEEE Asia-Pacific Microwave Conference*, pp. 1–4.

A. Italia, F. Carrara, A. Scuderi, et al., 2008, "Radio-Frequency Front-End for 5GHz Wireless Local Area Network Transceivers," *IET Circuits, Devices and Systems*, vol. 2, pp. 439–450.

A. Italia, E. Ragonese, L. La Paglia, and G. Palmisano, 2004, "A 5-GHz Silicon Bipolar Radio Transceiver Front-End," *Proceedings of the IEEE Bipolar/BiCMOS Circuits and Technol. Meeting*, pp. 120–123.

K. Jain, M. Klosner, M. Zemel, and S. Raghunandan, 2005, "Flexible Electronics and Displays: High-Resolution, Roll-to-Roll, Projection Lithography and Photoablation Processing Technologies for High-Throughput Production," *Proceedings of the IEEE*, vol. 93, pp. 1500–1510.

H. L. Kao, A. Chin, C. C. Huang, et al., 2005, "Low Noise and High Gain RF MOSFETs on Plastic Substrates," *IEEE International Microwave Symposium Digest*, pp. 295–298.

H.-T. Kim, L. Kai, R. C. Frye, et al., 2009, "Design of Compact Power Divider Using Integrated Passive Device (IPD) Technology," *Proceedings of the IEEE Electronic Components and Technology Conference*, pp. 1894–1899.

Z. Ma, and L. Su, 2009, "Will Future RFIC Be Flexible?" *Proceedings of the IEEE Wireless and Microwave Technology Conference*, pp. 1–5.

S. Ravesi, S. Alessandrino, A. Bassi, et al., 2008,"Materials and Processing Issues for the Manufacturing of Integrated Passive and Active Devices on Flexible Substrates," *Proceedings of the Flexible Electronics and Displays Conference and Exhibition*, pp. 1–3.

W. H. Teh, G. Lihui, R. Kumar, and D.-L. Kwong, 2005, "200-mm Wafer-Scale Transfer of 0.18-μm Dual-Damascene Cu/SiO$_2$ Interconnection System to Plastic Substrates," *IEEE Electron Device Letters*, vol. 26.

M. Zargari, D. K. Su, C. P. Yue, et al., 2002, "A 5-GHz CMOS Transceiver for IEEE 802.11a Wireless LAN Systems," *IEEE Journal of Solid-State Circuits*, vol. 37, pp. 1688–1694.

Glossary

AM: amplitude modulation
AMD: arithmetic mean distance
AMSD: arithmetic mean square distance
BEOL: back end of line
BiCMOS: Bipolar complementary metal oxide semiconductor
BJT: bipolar junction transistor
CMOS: complementary metal oxide semiconductor
CMP: chemical mechanical planarization
d_{in}**:** inner diameter
DUT: device under test
EM: electromagnetic
ETSI: European Telecommunications Standards Institute
FCC: Federal Communication Commission
f_{IM}**:** image frequency
f_{LO}**:** local oscillator frequency
f_{max}**:** maximum oscillation frequency
FOM: figure of merit
f_{SR}**:** self-resonance frequency
f_T**:** current gain cutoff frequency
GMD: geometric mean distance
GSG: ground-signal-ground
HBT: heterojunction bipolar transistor

IC: integrated circuit

IF (f_{IF}): intermediate frequency

IL: insertion loss

IPD: integrated passive device

IR: image rejection

IRF: image-reject filter

IRR: image-rejection ratio

ISM: industrial, scientific, and medical

ISS: impedance standard substrate

k: magnetic coupling factor

L: inductance

L_{DC}: low-frequency inductance

LNA: low-noise amplifier

LO: local oscillator

LRM: line-reflect-match

LRR: long-range radar

LRRM: line-reflect-reflect-match

MAG: maximum available gain

MCM: multichip module

MEMS: microelectromechanical systems

MIM: metal-insulator-metal

mm-wave: millimeter wave

MOSFET: metal oxide semiconductor field-effect transistor

n: number of turns

PA: power amplifier

PEN: polyethylene naphthalate

PG: pulse generator

PGS: patterned ground shield

PLL: phase-locked loop

PM: phase modulation

PSD: power spectral density

Q: quality factor

Q_{MAX}: maximum quality factor

R2R: roll-to-roll

RF: radio frequency

RFID: radio frequency identification

Rms: root mean square

RX: receiver

s: spacing

SiP: system-in-package

SoC: system-on-chip

SOI: silicon on insulator

SOLR: short-open-load-reciprocal

SOLT: short-open-load-through

SRR: short-range radar

t: thickness

TCR: transformer characteristic resistance

t_{ox}: oxide thickness

T_{PR}: pulse repetition interval

T_{pulse}: pulse width

TRL: through-reflect-line

TX: transmitter

UWB: ultra-wideband

VCO: voltage-controlled oscillator

VLSI: very large-scale integration

VNA: vector network analyzer

w: width

W2W: wafer-to-wafer

WLAN: wireless local area network

ε_{ox}: oxide dielectric permittivity

ε_{Si}: silicon dielectric permittivity

ρ_{Si}: silicon resistivity

Index